SUPPORTABILITY OF COMPOSITE AIRFRAMES

Proceedings of a Workshop Supportability of Composite Airframes on 3rd and 4th August 1987. Sponsored by US Air Force European Office of Aerospace Research and Development and Paisley College of Technology

SUPPORTABILITY OF COMPOSITE AIRFRAMES

Edited by

I. H. MARSHALL

Department of Mechanical and Production Engineering,
Paisley College of Technology, Scotland, UK

and

E. DEMUTS

US Air Force Wright Aeronautical Laboratories,
Wright-Patterson Air Force Base, Ohio, USA

Reprinted from the journal
Composite Structures
Vol. 10, No. 1

ELSEVIER APPLIED SCIENCE
LONDON and NEW YORK

ELSEVIER APPLIED SCIENCE PUBLISHERS LTD
Crown House, Linton Road, Barking, Essex IG11 8JU, England

Sole Distributor in the USA and Canada
ELSEVIER SCIENCE PUBLISHING CO., INC.
52 Vanderbilt Avenue, New York, NY 10017, USA

WITH 14 TABLES AND 54 ILLUSTRATIONS

British Library Cataloguing in Publication Data

Supportability of composite airframes.
1. Airframes. Composite materials
I. Marshall, I. H. II. Demuts, E.
III. Composite structures
629'.134'31

ISBN 1-85166-258-8

Library of Congress CIP data applied for

Supportability of composite airframes/edited by I. H. Marshall and E. Demuts.
p. cm.
'Reprinted from the journal, Composite structures, vol. 10, no. 1.'
Papers presented at a workshop held Aug. 3–4, 1987, in Glasgow, Scotland, sponsored by the United States Air Force European Office of Aerospace Research and Development and Paisley College of Technology.
Bibliography: p.
Includes index.
ISBN 1-85166-258-8
1. Airframes—Congresses. 2. Composite construction—Congresses. 3. Composite materials—Congresses. I. Marshall, I. H. (Ian H.). II. Demuts, E. III. United States. Air Force. European Office of Aerospace Research and Development. IV. Paisley College of Technology.
TL671.6.S84 1988
629.134'31—dc 19

88-18856
CIP

Typeset in Great Britain by Keyset Composition, Colchester
Printed in Great Britain by Galliard (Printers) Ltd, Great Yarmouth

Foreword

These proceedings are a record of a workshop entitled 'Supportability of Composite Air-Frame Structures' held in Glasgow, Scotland in August 1987. The workshop was jointly sponsored by the US Air Force European Office of Aerospace Research and Development and Paisley College of Technology.

The US Air Force promoted and supported the workshop because of the high level of emphasis given to reliability and maintainability (R&M) of aerospace systems. Air Force leadership is totally committed to increasing combat availability and capability through enhanced R&M. R&M are now primary objectives throughout the entire acquisition process, from conceptual development to design, production and employment. Performance is not enough. If you can't keep them flying, what good are they? Equipment must be designed with R&M as an essential requirement, equal to performance, cost and schedule, if operational requirements are to be met. R&M cannot be added later. It must be an integral part of the design process. All acquisition contracts should now be written to meet operational R&M needs.

This workshop provided a critical international forum for defining and accessing supportability of composite airframe structures, a key R&M aspect. Many thanks to the participants for sharing their views with us.

Special thanks are due to the conference organizers: Dr Ian Marshall of Paisley College of Technology and Edvins Demuts of the US Air Force Wright Aeronautical Laboratories. Due to unceasing efforts of these personnel, we attained a truly international workshop with participation from seven different countries.

Lt Col. James G. R. Hansen
Chief, Structures and Structural Materials
European Office of Aerospace Research and Development

Introduction

Herein are contained the papers presented at the Workshop 'Supportability of Composite Airframe Structures' held at the Grosvenor Hotel, Glasgow on the 3rd and 4th August 1987. The Workshop was jointly sponsored by the USAF European Office of Aerospace Research and Development and Paisley College of Technology.

This was indeed a unique event bringing together specialists in composite airframe structures from the United States, United Kingdom, Netherlands, Israel, Sweden, Australia and West Germany, each from an organisation deeply committed to the Workshop theme. Although in itself the bringing together of such diverse specialists spanning a wide range of military and civil aircraft organisations was an unprecedented event, the free and honest exchange of information, experience and philosophies made it doubly so.

Worldwide, vast sums of money and manpower are assigned to the 'supportability' of composite airframe structures, ranging from large spare parts inventories to dedicated maintenance personnel, from repair specialists to inspection equipment and so on. Much of this can be ascribed to past mistakes which may have compromised the availability of aircraft for full operational duties.

Indeed, it is only when the manifold connotations of the term 'supportability' are contemplated that the vast implications of material and manpower resources become apparent. Since it is envisaged that the future of both military and civil aircraft will depend largely on the efficient use of composites, it is surely fitting that the overall concept of supportability is addressed at this time in order to obtain a coherent worldwide strategy.

The objective of the workshop was to collect and digest the views on the supportability issue from a representative international gathering of specialists. To each of the contributors to this Workshop, the term 'supportability' had many meanings and priorities. However, it was reassuring to note a central theme and common commitment to seriously tackling the problem. It was agreed unanimously that only by freely exchanging past experiences and present policies could a meaningful strategy be devised. The term 'supportability' almost irritates the individual when attempting to write a concise definition of its global meaning. However, it should be remembered that it takes an irritant to produce a pearl.

Before attempting such a definition, it must be emphasised that the term cannot be restricted to simply maintainability and repairability. In so doing an imprecise and misleading limitation is imposed. Clearly a number of factors influence the choice of a suitable definition with specific regard to the airframe structure, these being reliability, maintainability, assessability, repairability and producibility.

Thereafter it is apparent that each of these terms has a multiplicity of subfactors, many of which are inter-connecting. Non-destructive evaluation (NDE) has implications for all of these headings, likewise damage tolerance and so on. However, it is obvious that many of the technical aspects relating to these headings have yet to be adequately and universally addressed, e.g. it is clear that no unified damage tolerance methodology exists between organisations in the same country, far less worldwide. There is also a great need for a simple NDE method able to detect delaminations in skins and substructures. Indeed many would say that, in spite of vast advances in the past decade, the whole subject of NDE is still in its infancy with respect to composite structures. Notwithstanding such limitations, it should also be borne in mind that escalating aircraft replacement costs, particularly military aircraft, are forcing original operational design envelopes to be extended, sometimes by substantial margins. However, since most experience in the use of primary composite airframe structures is in military aircraft, this has positive benefits with regard to a greater understanding of supportability aspects. Since many specific technical and logistical aspects are covered by the individual contributors to this volume, it would seem fitting to restrict these introductory comments to more general aspects.

Whatever the specific technical branches of supportability are, there is no doubt that its roots should lie at the design stage. Only by fully appreciating its manifold implications at that time can real progress be made which will undoubtedly lead to both substantial cost and manpower savings and also more efficient composite airframe structures.

Education must also play a leading role, not just of the designer and other high-level technical personnel, but others such as maintenance and stores personnel. In the latter case, these individuals must be made to appreciate that composite structures are inherently different from their metallic equivalents. It is clear that a considerable portion of the accidental damage to such structures is caused, not by runway debris or bird impact during takeoff and landing, but by transportation from stores to aircraft or by accidental dropping of hand tools during maintenance.

In summary, a greater co-operation between engineering and logistical disciplines would make sizeable inroads into the overall supportability problem.

Finally, we may grasp a very difficult and thorny nettle and give our definition of supportability as applied to composite airframe structures. This would be 'The total ability of a composite airframe structure to meet the aircraft design life for all operational conditions, both present and envisaged'.

The Editors would like to express their appreciation of the efforts of Lt Col. James G. R. Hansen in promoting this Workshop and assisting in its successful completion and also Robert M. Bader, Assistant Chief, Structures Division, Flight Dynamics Laboratory, WPAFB, Ohio, USA without whose active and enthusiastic support the workshop would not have been possible. Thanks are also due to the Workshop Secretary, Miss Alison Shedden, for her tireless efforts in organising the Workshop and generally ensuring its smooth running.

As always, a special thanks to our respective families for their fore-bearance and support, both before and during the Workshop.

I. H. Marshall
E. Demuts

Contents

Composite Structures **10** (1988) 1–15

Non-Destructive Test Analysis and Life and Residual Strength Prediction of Composite Aircraft Structures

I. R. Farrow & J. B. Young

Aircraft Design Department, College of Aeronautics, Cranfield Institute of Technology, Cranfield, Bedford, UK

ABSTRACT

Supportability is defined in this paper as the ability to assess and reassess the effect of change in operational use on fatigue damage accumulation. To achieve this a rationalised approach and cumulative damage model are proposed to relate Non-Destructive Evaluation measurements to life and residual strength of advanced composite aircraft structures.

1 INTRODUCTION

This paper is concerned primarily with Fatigue Damage of advanced composite aircraft structure and as such presents a diverse aspect to the mainstream concern of impact damage in this 'Workshop'. Many of the concepts used may however be applicable to damage accrued by means other than fatigue. At current aircraft design stress levels and life requirements the fatigue of advanced fibre composites is a secondary design consideration. Design to static requirements can usually cover fatigue requirements. As such the present solution is to use a Safe Life design philosophy. However, in doing so there are implications regarding the 'design supportability' of aircraft composite components:

Question: How safe is the safe life:
if operational life is extended beyond original design?
if operational usage is outside the original design envelope?
if localised loading is discovered to be greater than design?
if damage is detected which is not covered in design?
Answer: Less safe!

Composite Structures 0263-8223/88/$03·50

Until a capability is developed to predict the effect of variations in operational use, then the supportability in terms of design reassessment will be severely restricted. Although conservative approximation may be a good design virtue it does not lead to an understanding of the problem. Introducing too many safety factors may easily lead to an unbalanced technical solution.

For future applications of advanced composite materials it might be expected that more efficient use will be made of the material at higher stress levels or longer fatigue lives so that the ability to accurately predict fatigue life becomes of greater importance. Further, in solving the impact damage tolerance problem of advanced composite we may well find our fibre resin system has become less fatigue tolerant.

To achieve 'design supportability' the current Safe Life philosophy must give way to a Fail Safe damage tolerant philosophy. We must be able to estimate the fatigue damage accumulation and residual strength at any intermediate stage in the operational life and the duration to the failure damage state without the need for spectrum fatigue tests for every conceivable variation of service use. With these complex tasks to be carried out we should not lose sight of the need for a simple method of analysis based on simple performance data since many schemes may be considered with consequent repetition of calculations.

2 FAIL SAFE DAMAGE TOLERANT DESIGN

Damage tolerant performance of advanced composite materials relies on quantitative Non Destructive Evaluation (NDE) methods to provide Characteristic Damage Parameters (CDP). The cumulative damage process in fibre composite materials consists of various competing damage modes which lead to a change in state of the material and redistribution of stresses in the laminate (Fig. 1). It is unrealistic to attempt to measure the effect of each individual damage event on the properties of the laminate in question under all possible loading conditions. Rather, the aim is to establish a means of measuring the collective damage effect.

There are many indications of fatigue damage in composites which can be exploited to develop in-service inspection schemes. Examples of NDE techniques which are potentially useful for quantitative analysis are:[1–7]

Ultrasonics
Acoustic emission
Thermal imaging
X-radiography
Dynamic mechanical analysis: resonant frequency, damping coefficient . . .

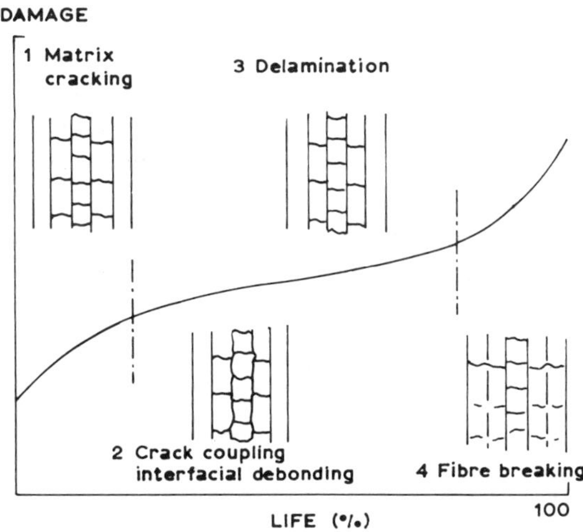

Fig. 1. Damage modes observed in composite laminates.[34] (Reifsnider, Stinchcomb).

However, each method is limited to revealing certain damage modes in certain types of material and structure, all with varying on site capability. No technique is yet fully quantifiable. It is conceivable that no single technique will prove capable of giving a complete picture of the damage state. Rather the use of two or more NDE techniques is envisaged, particularly useful if they overlap to provide a more confident assessment. If we consider the situation where monitoring shows that damage has been accrued and then appears to stop but still results in unexpected failure at a later stage in life, then it must be that the damage has not in fact stopped growing it has simply taken another of the possible damage modes which is not detected by the selected NDE technique.

In this work the objective has not been to develop and verify any particular NDE technique but rather to develop a rationalised philosophy and cumulative damage rule which could readily be applied to future quantitative NDE–CDP databases. To this end, effective dynamic stiffness has been measured during fatigue tests as crude but easily obtainable ND data to develop the proposed prediction technique.[8–14]

Useful results have already been obtained for carbon fibre cross ply and quasi-isotropic laminate notched specimen in this research. However, use of absolute values of initial dynamic stiffness as a damage parameter has shown insufficient correlation to endurance and residual strength for the sample sizes used. As a first approach the change in stiffness from an assumed initial condition has been used adjusted to a low percentile statistic to simulate the highest expected rate of degradation.

3 FATIGUE DAMAGE ANALYSIS

It is the authors' view that there is no generally applicable fatigue analysis approach available for fibre composite structural design. Contemporary methods such as Miners and Linear Elastic Fracture mechanics have been unsuccessful,[15–20] and many methods currently being developed specifically for composite materials remain incomplete and restricted in their field of application. Detailed discussion of available methods is outside the intended scope of this presentation and is covered in part by the references.[10,12,13,14,17,21–41]

The objectives of any fatigue damage analysis method are to provide a general application dependent on a small, readily obtainable database, avoiding theory which becomes too involved or models which become too contrived. With this in mind, part of this work has involved setting out to develop a prediction technique to meet these objectives, with simplicity for initial design application and in-service design support being the main aim.

The first task has been to plan a rationalised approach to the main problems that confront us in the fatigue analysis of advanced composite structures. The viewpoint taken is summarised in the following sections.

3.1 Macromechanic

Problem: The fatigue process can be extremely complex involving many different and interacting failure mechanisms. The damage can consist of fibre breakage, matrix cracking, fibre pull-out, delamination and so forth (Fig. 2). The damage usually accumulates as a complex array of cracks throughout the material.[35,42–45]

Approach: Provide a practical engineering solution for the structural designer who will be interested only in those aspects that manifest themselves at a macroscopic level; avoiding attempts to solve complex micromechanic mechanisms of the fibre composite damage accumulation.

3.2 Degradation

Problem: The complex array of damage in the material causes a degradation of material properties, and effective failure can occur well before catastrophic fracture (Figs 3–5).[8–13]

Approach: Make use of a measurable manifestation of macroscopic damage as Characteristic Damage Parameters. Ensure that failure is concisely defined as a critical level of degraded property—measured as a damage parameter—for a given probability of earliest occurrence (the level is assumed to describe a certain critical macroscopic damage state). The duration to achieve this level is then the fatigue life (Fig. 6).

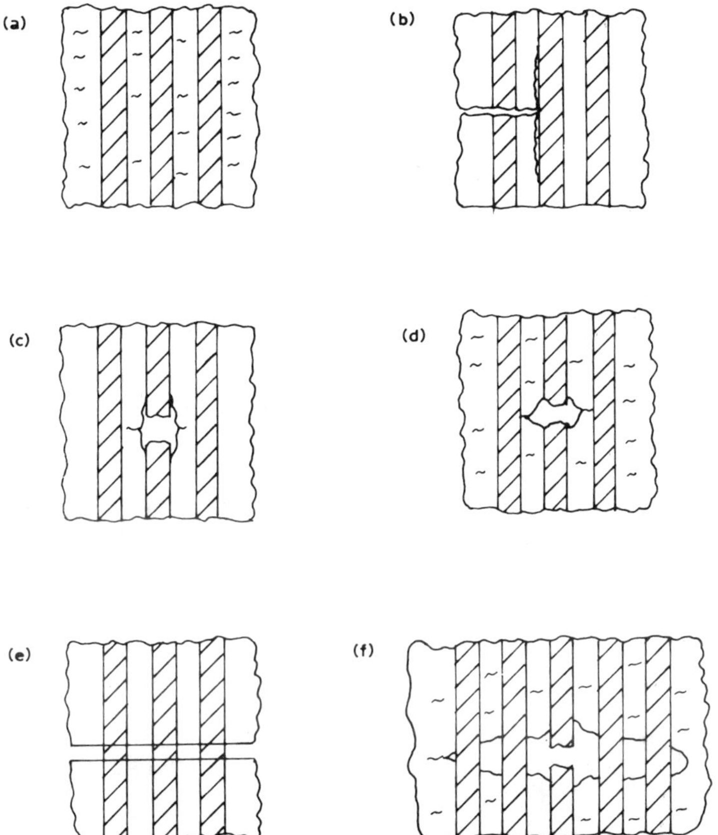

Fig. 2. Damage modes:[45] (a), dispersed cracks confined to matrix only; (b), cracks growing by breaking fibres leading to interface failure; (c), fibre break causing interfacial debonding; (d), fibre break increasing matrix crack; (e), fibres bridging matrix crack; (f), combination.

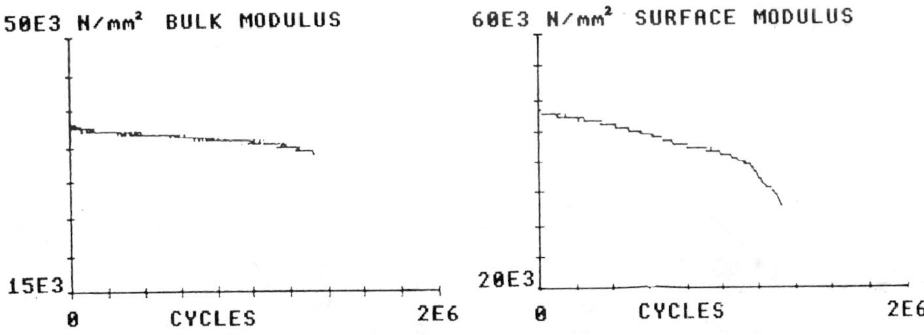

Fig. 3. Typical stiffness degradation of quasi-isotropic fastener specimen under R0.1 constant amplitude stress cycling.

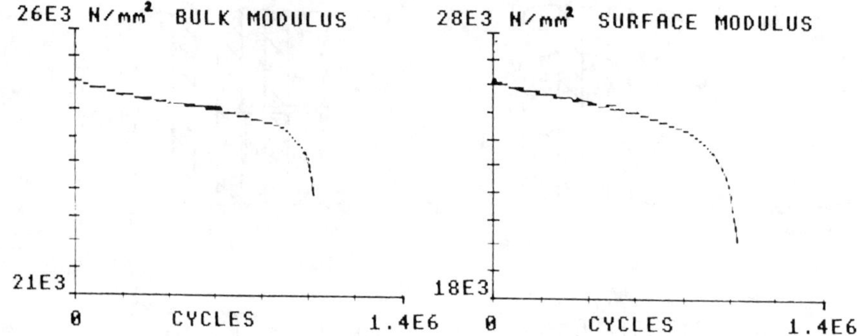

Fig. 4. Typical stiffness degradation of cross ply fastener specimen under R0.1 constant amplitude stress cycling. Bulk means actuator measurement of strain (including grip effect); surface means extensometer measurement of strain.

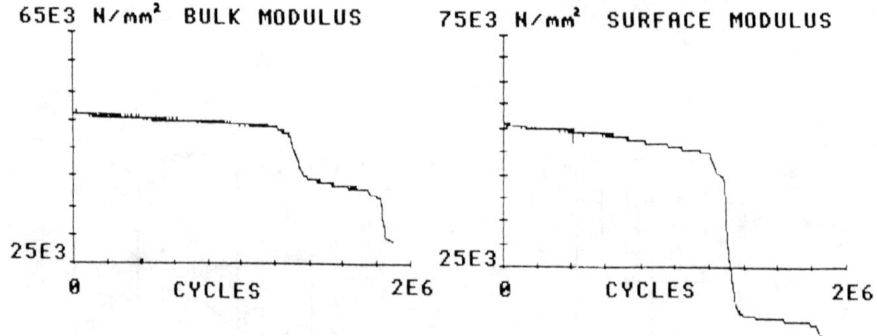

Fig. 5. Stiffness degradation of quasi-isotropic fastener specimen under R-1 constant amplitude stress cycling at 50% UTS peak stress.

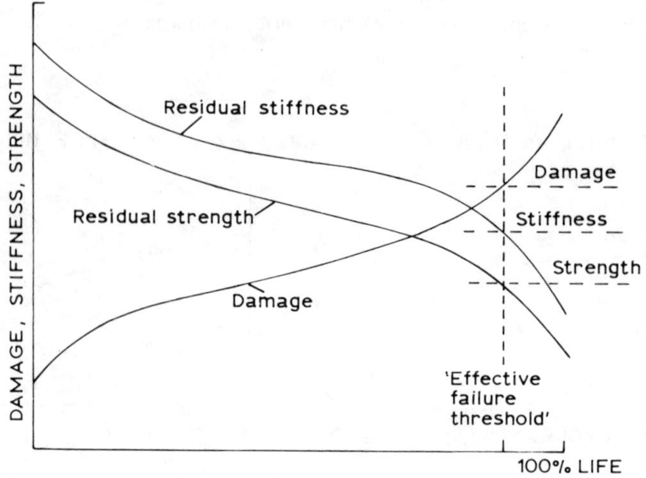

Fig. 6. Schematic presentation of effective failure definition.

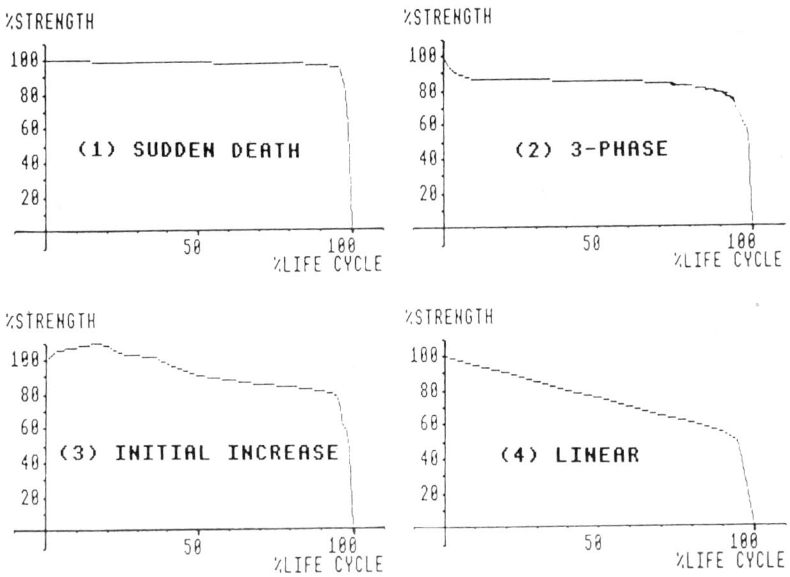

Fig. 7. Residual strength trends.

Investigate correlation between the chosen characteristic damage parameters and residual strength. Deduce residual strength to the required degree of confidence level from the corresponding level of damage parameter. The resulting residual strength envelope may be used as a 'design window'. This removes the need to model degradation and the necessity of assuming a direct relationship between initial ultimate strength distribution and the final fatigue failure distribution (Fig. 7).

3.3 Probabilistic

Problem: Composite materials have an inherent property variability. This is due to the very nature of the material:

Fibres/resin/plies/laminate/bond/notch

Each item and its interface with the others provides a compounding system of variability (Figs 8 and 9).

To further compound this problem we find that advanced composites are particularly susceptible to damage by singular high loads which makes the application of a proof stress impractical; leaving no means of screening out the weak components in the tail end of the strength distribution.

Approach: Use statistical presentation of any damage parameters, in particular the extreme value statistic including the tail end of the distribution

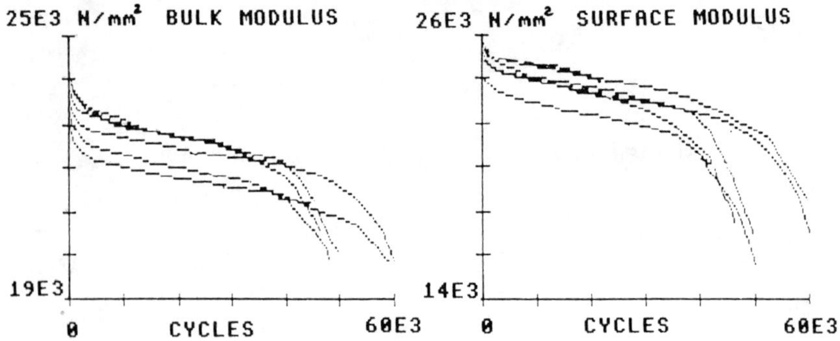

Fig. 8. Stiffness degradation curves for sample set of six cross ply fastener specimen under R-1 constant amplitude stress cycling at 40% UTS peak stress.

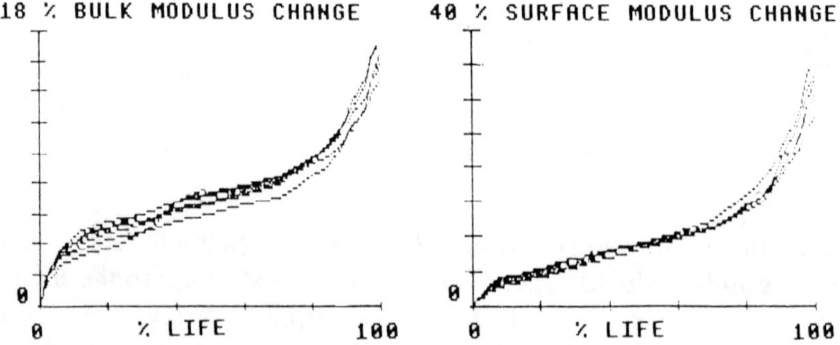

Fig. 9. Percentage change in stiffness derived from the results in Fig. 8.

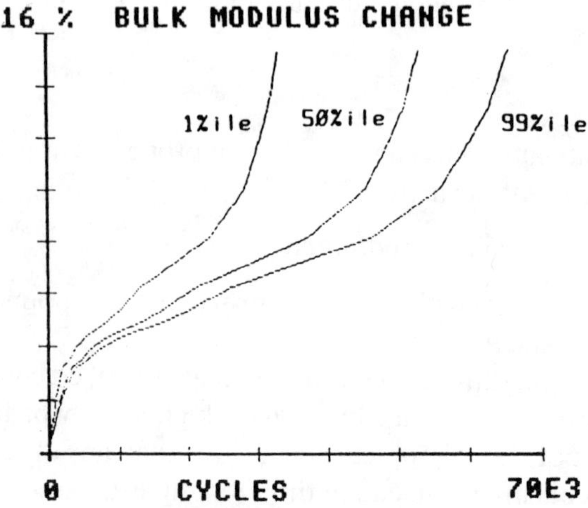

Fig. 10. Weibull percentile degradation curves derived from Fig. 9 (assumed 'standard' initial condition as first approximation).

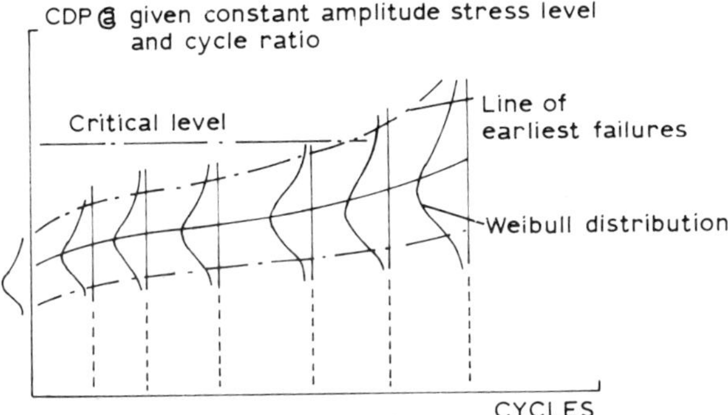

Fig. 11. Schematic presentation of Weibull percentile degradation curves for CDP using absolute initial value.

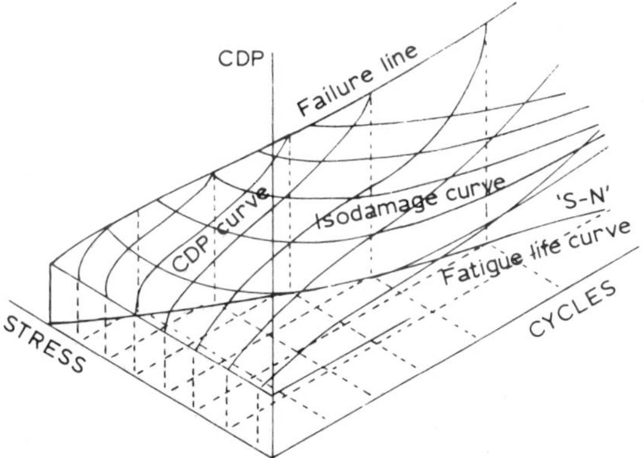

Fig. 12. Schematic presentation of CDP database derived from Weibull extreme value statistic of degradation curves at selected constant amplitude stress levels.

which describes the earliest failures. (The most commonly preferred distribution being that of Weibull.) (Figs 10 and 11).

3.4 Damage accumulation

Problem: Non-linear damage accumulation with the possibility of significant sequence effect from a deterministic/random load spectrum. This makes the task of extrapolating the effect of changes in load spectra extremely difficult.[18,46–54]

Approach: Develop non-linear damage accumulation model where damage accumulated depends on current damage state allowing for sequence effect. Avoid further idealisation of already idealised load spectra. Maintain a level of representation of load sequence by applying the idealised spectra directly to the cumulative model by a predefined iterative procedure. (As a first approach only the increasing positive and decreasing negative parts of the cycles will be considered as damage effective.) Predict the limit of safe damage state rather than ultimate failure.

Make use of a base data-set consisting of constant amplitude fatigue and residual strength with a 'map' of degrading damage parameters for the laminate under investigation (Fig. 12).

A demonstration of the proposed damage accumulation model is given in Appendix B.

4 DISCUSSION

It is envisaged that the structure under investigation will be mapped into various structurally significant areas categorised into structural types each described by a set of 'damage parameter' databases provided by the selected NDE techniques. Only after this approach is verified and generalised can we move on to predict the functional forms of the degradation and isodamage curves for various derivative laminates.

This research is continuing with first prediction results due by the end of 1987. Details of materials, specimen and laminate types investigated are given in Appendix A.

REFERENCES

1. Chang, F. H., Gordon, D. E. & Gardener, A. H., *A Study of Fatigue Damage in Composites by Non-Destructive Testing Techniques*. ASTM STP 636, 1977, pp. 57–72.
2. Hamstad, M., A review: acoustic emission, a tool for composite material studies. *Experimental Mechanics*, **26** (Mar. 1986) 7–13.
3. Harris, B., Assessment of structural integrity of composites by non-destructive methods. *AGARD. LS-124*, 1982.
4. Marcus, L. A., Stinchcomb, W. W., Measurement of fatigue damage in composite materials. *Experimental Mechanics*, **15** (Feb. 1975) 55–60.
5. Marr, G. R., Barrowcliffe, R. P. & Curtis, A. R. The non-destructive evaluation of composite bonded joints. *ICCS 3, Proc. Paisley*, 1985, pp. 502–10.
6. Whitcomb, I. D., *Thermographic Measurement of Fatigue Damage*. ASTM STP/674, 1979, pp. 502–16.
7. Whitcomb, J. O., *Experimental and Analytical Study of Fatigue Damage in Notched Graphite/Epoxy Laminates*. ASTM STP/723, 1981, pp. 49–63.
8. Camponeschi, E. I. & Stinchcomb, W. W., *Stiffness Reduction as an Indicator of Damage in Graphite/Epoxy Laminates*. ASTM STP/787, 1982, pp. 225–46.
9. Gerharz, J. J., Development of damage/inspection procedures for fibre

composites. *Royal Aeronautical Establishment Trans.*, Report No. 2045 (Aug. 1980) pp. 7.1–7.20.
10. Gottesman, T. & Mikulinsky, M., Influence of matrix degradation on composite properties in the fiber direction. *Engineering Fracture Mechanics*, **20**(4) (1984) 667–74.
11. Laws, N., Dvorack, G. J. & Hejazi, M., Stiffness changes in unidirectional composites caused by crack systems. *Mechanics of Materials*, **2** (1983).
12. Ogin, S. L., Smith, P. A. & Beaument, P. W. R., Matrix cracking and stiffness reduction during the fatigue of a (0/90)S GFRP laminate. *Composite Science and Technology*, **22**(1) (1985) 23–31.
13. Poursartip, A. & Beaument, P. W. R., A damage approach to the fatigue of composites. In *Mechanics of Composite Materials, Proc. Symp. IUTAM, Blacksburry*, VA, USA, 1983, pp. 449–56.
14. Reifsnider, K. L., The mechanics of fatigue in composite laminates. *Proc. Japan–USA Conf. on Composite Materials*, Tokyo, 1981.
15. Achard, R. T. & Richardson, M. D., Load spectrum effects on bonded composite joints. *AFFDL TR-77-29*. 770400, 1977.
16. Arendts, F. J., Sippel, K. O. & Weisberger, D., Constant amplitude and flight by flight tests on CFRP specimen. *AGARD CP 288*, 50th meeting, Athens, 1980.
17. Gordon, B. & Whitehead, R. S., A critical review of life prediction methods for composite structures. BAE Report No. SON(P) 179. Feb, 1979.
18. Heath Smith, J. R., Fatigue of structural elements in carbon fibre composites—present indications and future research. Royal Aeronautical Establishment, Report, TR-79085, July 1979.
19. Kanninen, M. F., Rybicki, E. F. & Brinson, H. F., A critical look at current applications of fracture mechanics to the failure of fibre-reinforced composites. *Composites J.*, **8** (Jan. 1977) 17–22.
20. Kanninen, M. F., Rybicki, E. F. & Griffith, W. I., *Preliminary Development of a Fundamental Analysis Model for Crack Growth in a Fiber Reinforced Composite Material.* ASTM STP/617, 1977, pp. 53–69.
21. Amijima, S., Tanimoto, T. & Matsuoka, I., A study on fatigue life estimation of FRP under random loading. In *Progress in Science and Engineering of Composites: Proc. 4th International Conference on Composite Materials, Tokyo, Japan*, **1** (1982) 701–8.
22. Chou, P. C., *Degradation and Sudden Death Models of Fatigue of Graphite/Epoxy Composites*. ASTM STP/674, 1979, pp. 431–54.
23. De Iorio, A., Mignosi, S. & Schiavon, M., Fatigue in CFRP under variable amplitude loading. In *Testing, Evaluation and Quality Control of Composites, Proc. The International Conference*, Guildford, Surrey, England, 1983, pp. 314–23.
24. Giancarlo Caprino, On the prediction of residual strength for notched laminates. *J. Materials Science*, **18** (1983) 2269–73.
25. Hahn, H. I., *Fatigue Behaviour and Life Prediction of Composite Laminates*. ASTM STP/674, 1979.
26. Halpin, J. C., Jerina, K. L. & Johnson, T. A., *Characterisation of Composites for the Purpose of Reliability Evaluation.* ASTM STP/521, 1973.
27. Hashin, Z., Cumulative damage theory for composite materials: residual life and residual strength methods. *Composites Science and Technology*, **23** (1985) 1–19.

28. Kulkarni, S. V., McLaughlin, P. V., Pipes, R. B. & Rosen, B. W., *Fatigue of Notched Fiber Composite Laminates: Analytical and Experimental Evaluation.* ASTM STP/617, 1977, pp. 70–92.
29. Lauritis, K. N., Ryder, J. T. & Pettit, D. E., Advanced residual strength degradation rate modelling for advanced composite structures (Vol. 3). *AFWAL TR-79-3095.*
30. McLaughlin, P. V., Kulkarni, S. V., Huang, S. N., Walter, B. & Rosen, B. W., Fatigue of notched fiber composite laminates. Part 1: Analytical Model. *NASA CR-132-747*, Mar. 1975.
31. Pipes, R. B., Kulkarni, S. V. & McLaughlin, P. V., Fatigue damage in notched composite laminates. *Materials Science and Engineering*, **30** (1977) 113–200.
32. Radhakrishan, K., Fatigue and reliability evaluation of unnotched carbon epoxy laminates. *J. Composite Materials*, **18** (Jan. 1984) 21–31.
33. Ratwani, M. M. & Kan, H. P., Compression fatigue analysis of fiber composites. *J. Aircraft*, **18**(6) American Institute of Aeronautics and Astronautics, Report No. 80-0707R, 1980, pp. 458–62.
34. Reifsnider, K. L., Henneke, E. G., Stinchcomb, W. W. & Duke, J. C., Damage mechanics and NDE of composite laminates. In *Mechanics of Composite Materials, Proc. Symp. IUTAM; Blacksburry*, VA, USA, 1983, pp. 399–420.
35. Reifsnider, K. L. & Highsmith, A., Characteristic damage states: a new approach to representing fatigue damage in composite laminates. *Materials, Experimentation and Design in Fatigue, Proc.*, Westburry House, Guildford, Surrey, UK, 1981, pp. 246–60.
36. Reifsnider, K. L., Stinchcomb, W. W., Henneke, E. G. & Duke, J. C., Fatigue damage—strength relationships in composite laminates. *AFWAL TR-3084*, **1** Sept. 1983.
37. Salkind, M. J., *Fatigue of Composites*. ASTM STP/497, 1972, pp. 143–69.
38. Whitney, J. M., *Fatigue Characterisation of Composite Materials.* ASTM STP/723, 1981, pp. 133–51.
39. Whitney, J. M., *A Residual Strength Degradation Model for Competing Failure Modes.* ASTM STP-813, 1983, pp. 225–45.
40. Wnuk, M. P. & Kriz, R. O., COM model of damage accumulation in laminated composites. *International J. Fracture*, **28** (1985) 121–38.
41. Yang, J. N. & Du, S., An exploratory study into the fatigue of composites under spectrum loading. *J. Composite Materials*, **17** (Nov. 1983) 511–26.
42. Boniface, L. & Bader, M. G., Damage development in CFRP laminates under cyclic loading at constant and variable stress amplitudes. Surrey Univ., Dept. of Metallurgy and Materials Technology, Progress Report RAE. Farnborough ref: 2064/066/XR/MAT, 1983.
43. Gerharz, J. J., Mechanisms of fatigue damage and fatigue testing. *AGARD. LS-124*, 1982.
44. Harris, B., Fatigue and accumulation of damage in reinforced plastics. *Composites J.*, **8** (Oct. 1977) 214–20.
45. Talreja, R., Damage models for fatigue of composite materials. *Proc. Rino International Symp. on Materials Science Fatigue and Creep of Composite Materials*, Sept., 1982, pp. 137–53.
46. Badaliance, R., Dill, H. D. & Potter, J. N., *Effects of Spectrum Variations on Fatigue Life of Composites.* ASTM STP/787, 1982.

47. Bader, M. G. & Boniface, L., The performance of CFRP laminates subjected to complex fatigue loading programmes. *Fibre Reinforced Composites* (April 1984), pp. 7.1–7.11.
48. Bader, M. G. & Boniface, L., The assessment of fatigue damage in CFRP laminates. In *Testing, Evaluation and Quality Control of Composites, Proc. The International Conference*, Guildford, Surrey, England, 1983, pp. 66–75.
49. Broughtman, L. J. & Sahu, S., *A New Theory to Predict Cumulative Fatigue Damage in Fiberglass Reinforced Plastics*. ASTM STP/497, 1972.
50. Demuts, E., Fatigue spectrum sensitivity of composite joints—in aircraft structures. *Composites J.*, **14** (Jan. 1983) 27–33.
51. Schutz, D. & Gerharz, J. J., Fatigue strength of a fibre-reinforced material. *Composites J.*, **8** (Oct. 1977) 245–50.
52. Tanimoto, T., Amijima, S. & Ishikawa, H., Fatigue life and its reliability of FRP under multi-step loading. *Proc. Japan–USA Conf. on Composite Materials*, Tokyo, 1981.
53. Waddoups, M. E. & Halpin, J. C., The fracture and fatigue of composite structures. *Composites and Structures*, **4** (1974) 659–73.
54. Yang, J. N. & Jones, D. L., The effects of load sequence on statistical fatigue of composites. *AIAA J.*, **18**(12) (Dec. 1980).

APPENDIX A: DETAILS OF MATERIAL AND SPECIMEN TYPES UNDER INVESTIGATION

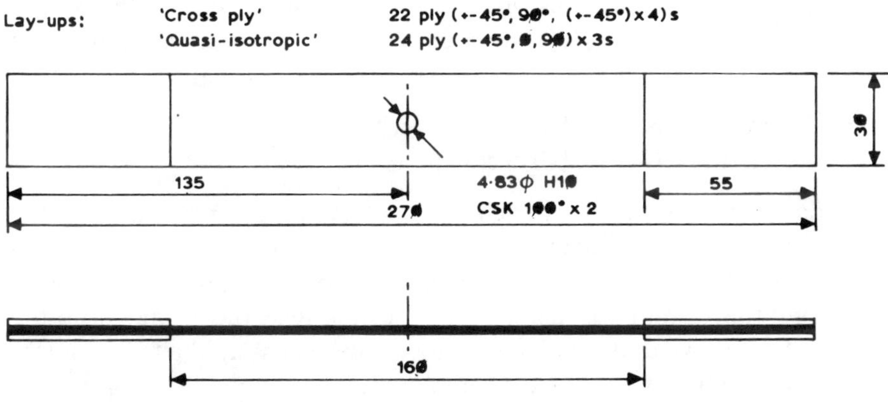

Fig. A1. Specimen details.

Fig. A1.—*contd.*

Ply drop-off specimen

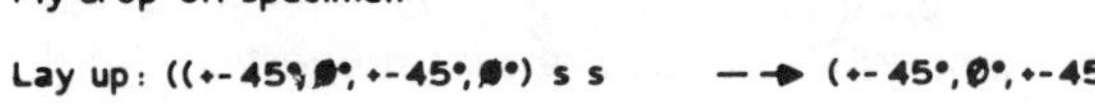

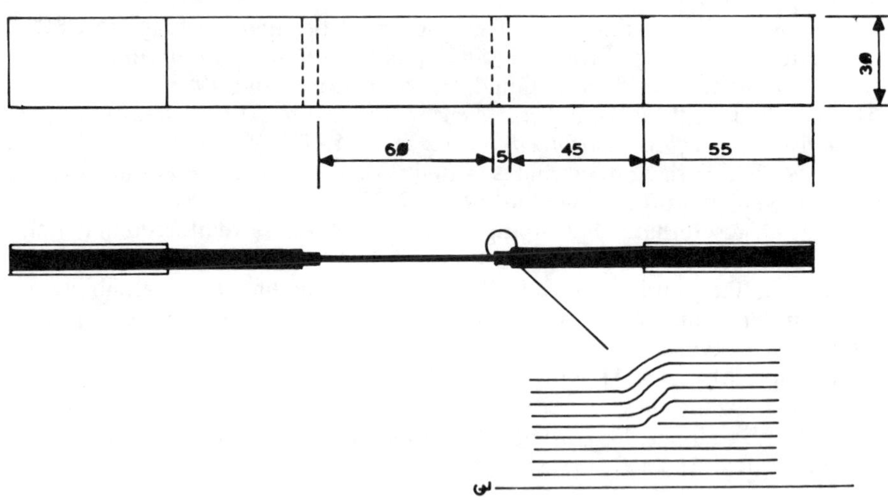

Fig. A1.—*contd.*

APPENDIX B: JOURNEY IN THE FATIGUE LIFE OF AN ADVANCED COMPOSITE STRUCTURAL ELEMENT

Consider two separate fatigue 'journeys' (see Fig. B1) involving two variable amplitude load histories consisting of blocks of cycles (individual cycles or blocks of pure random cycles could equally well be used). The only difference in the two load histories is the order in which the blocks are applied, the blocks themselves are identical in stress level and number of cycles.

e.g.: Block 'a' consists of Na cycles at Sa stress level
Block 'b' consists of Nb cycles at Sb stress level

The damage resulting from each block of cycles is represented by the increased damage level reached at the end of the block, i.e. the height of the shaded triangle. When each load block is completed the procedure is to track along the resulting isodamage level to the next block, move up the CDP database curve to the end of the block and so on until the failure line is reached or until the load history is completed.

It should be noted that load history 1 reaches the failure line at the end of the last block but load history 2 (which is merely history 1 reordered) does not reach the failure line and has a remaining residual strength which could be estimated from the corresponding residual strength envelope. The failure definition takes into account the current load level, i.e. failure occurs when

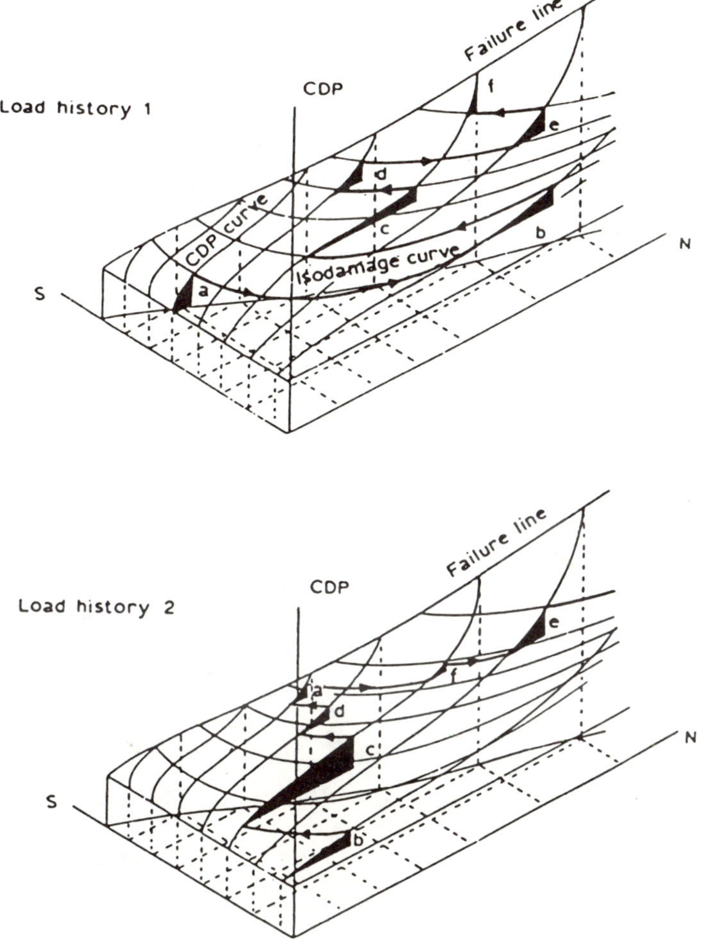

Fig. B1. Journey in the fatigue life of a composite structural element.

the damage level reaches the critical level for the current loading—indicated by the intersection of the CDP curve and failure line. The reciprocal of this is explained by considering residual strength as the characteristic damage parameter (which would be represented by a 'strength degradation surface') where failure occurs when the residual strength falls to the level of the applied load. The inference is that components under lower loading can accumulate a greater amount of damage before failure.

Composite Structures **10** (1988) 17–36

Supportability of Composite Airframes: An Integrated Logistic Viewpoint

W. Geier & J. Vilsmeier

Messerschmitt-Bölkow-Blohm GmbH, Helicopter and Military Aircraft Group, Postfach 80 11 60, D-8000 München, FRG

ABSTRACT

The aim of this paper is to highlight supportability of composite aircraft structure from an integrated logistic support point of view. Under this view, availability and life cycle costs become important parameters to assess the supportability of a system. Good supportability will be achieved by close cooperation between engineering and logistic disciplines and by performing a comprehensive reliability and maintainability programme. As an important element of supportability, repair procedures for composite material known to date do not yet adequately consider the logistic interests of the aircraft users.

1 DEFINITION OF SUPPORTABILITY

Supportability is the degree to which system design characteristics and planned logistics resources, including manpower, meet system peacetime operational and wartime utilization requirements (Figs 1–4). Important parameters used to measure the degree of supportability are availability and life cycle costs of a system in its operational environment.

Good supportability can be achieved by influencing the design, identifying the logistic support required by the system and by optimizing the support concept. Design to supportability can be achieved through the performance of an intensive reliability and maintainability (R & M) programme. Essential features of an R&M programme are design requirements, R & M analysis, design reviews and R & M demonstrations. A proposal of how to establish R & M programmes is provided by MIL STD

Composite Structures 0263-8223/88/$03·50

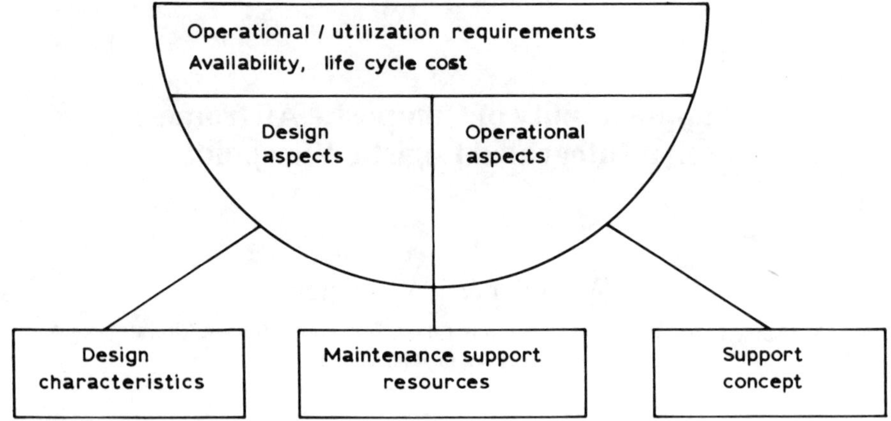

Fig. 1.

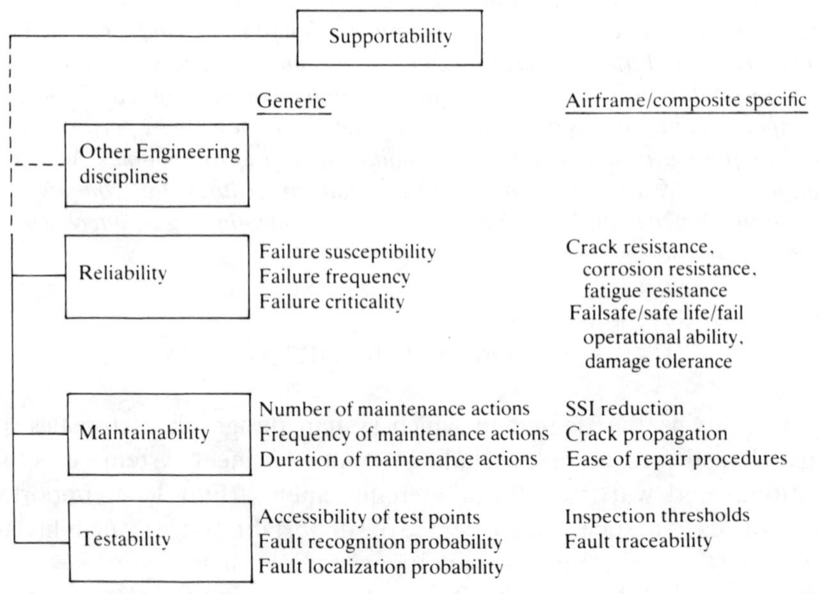

Fig. 2.

785 Reliability Programme Plan and MIL STD 470 Maintainability Programme Plan respectively. The logistic support required by the system will be identified and described by a detailed maintenance task analysis. A cost effective support concept will be developed by applying Optimum Repair Level Analysis. The complete process is called Logistic Support Analysis which is defined in MIL STD 1388.

Supportability
Generic
Airframe/composite specific
Support Personnel
Number
Skill level
Training requirements
Usage of detection equipment, performance of repairs
Support Equipment
Maintenance and handling equipment
Test and diagnostic equipment
Standardization
Vacuum bag, heated
Sensor technology (ultrasonic squirter etc.)
Commonality
Spares
Optimum repair level
Spares provisioning and supply
Modularization
Repair material
Document
Maintenance and repair manuals
Facilities
Infrastructure
Cooling facilities for repair kit storage (prepreg)
Vacuum cleaning facilities, exhaustors

Fig. 3.

Supportability
Generic
Specific
Allocation of maintenance tasks
Lines of maintenance (O, I, D)
Levels of maintenance
Tendency to de-emphasize maintenance
Large SSI's due to high integration of structural items
Deployment of maintenance sites
Transportation distances
Allocation of resources
Fewer maintenance sites
Combination manufacturing maintenance
Intensive usage of industrial repair sites
Support management
Organization
Procedures
Control

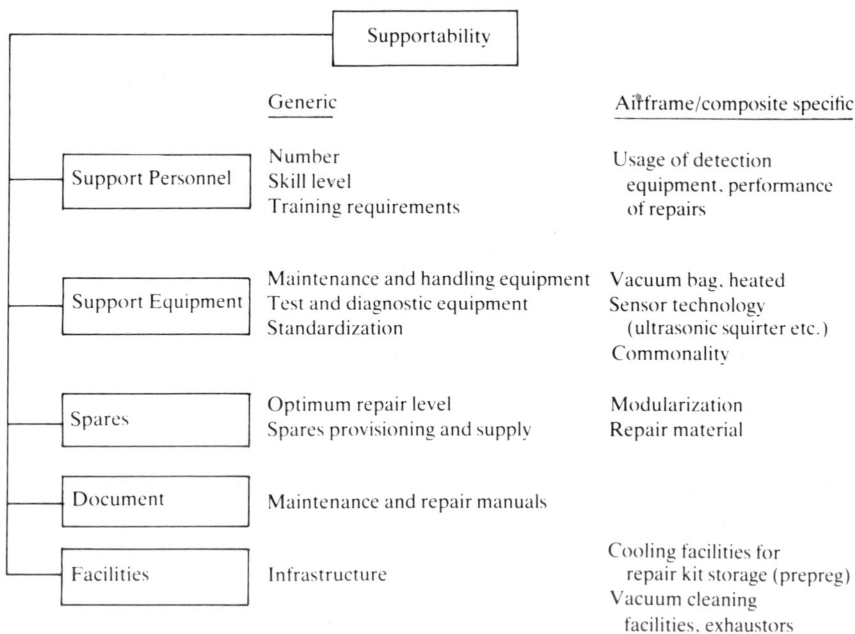

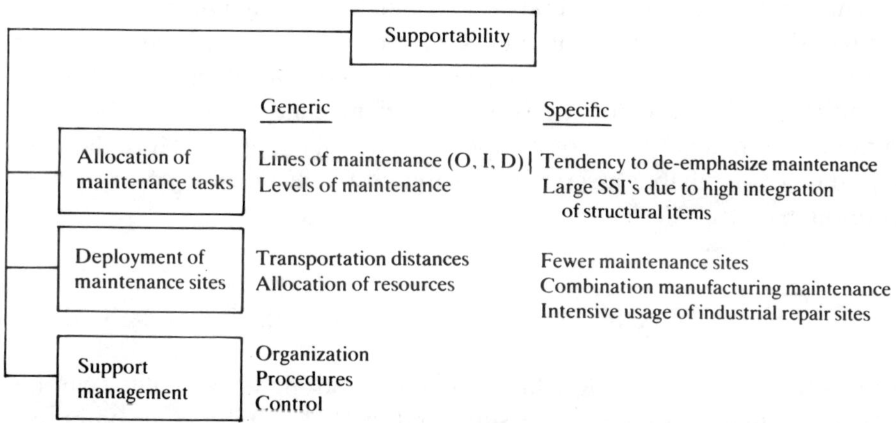

Fig. 4.

2 SUPPORTABILITY RELATED DISCIPLINES

Design to supportability can be achieved by close liaison between system engineering and R & M engineers. To identify and describe the logistic support required, close cooperation between R & M engineers and logistic disciplines for ground support equipment, training, material support and technical documentation is necessary (Fig. 5). Over the whole life cycle of a system, supportability, as part of Integrated Logistic Support, will interface with configuration control, design, repair and overhaul, user, and production. All these elements are embedded in a network which allows communication between each (Fig. 6).

3 SUPPORT EFFORT FOR AIRFRAMES

The airframe is only one of several systems of an aircraft. It is therefore important to consider the support effort in relation to other systems (Fig. 7). Considering helicopters and fixed wing aircraft, it turns out that the contribution of the airframe to the maintenance may come to 30%. About 50% of the airframe maintenance results from hardware failures (fasteners etc.). However, as a proportion of total maintenance costs the airframe contributes only 5%.

The question of whether or not there are cost-savings from the use of composite material cannot be conclusively answered due to the limited data available. Some examples of operational experience with composite components in F15 and F16 aircraft do not indicate a significant difference compared with metallic components with respect to mean time between failure (MTBF) (Table 1). Repair costs expressed in available mean time to repair (MTTR) data do not show any significant differences between components made of metal or composite material (Table 2). More data are required for a reliable assessment of the cost effectiveness of composite material in aircraft design.

4 RELIABILITY AND MAINTAINABILITY DESIGN FACTORS

Good supportability, i.e. low life cycle costs and high availability, can be achieved primarily by design to supportability. Reliability and maintainability are the most important disciplines that influence the design through the performance of an intensive R & M programme. One of the first tasks of such a programme is to establish design requirements or factors, and to ensure that these factors are met by the design. Design factors are mainly

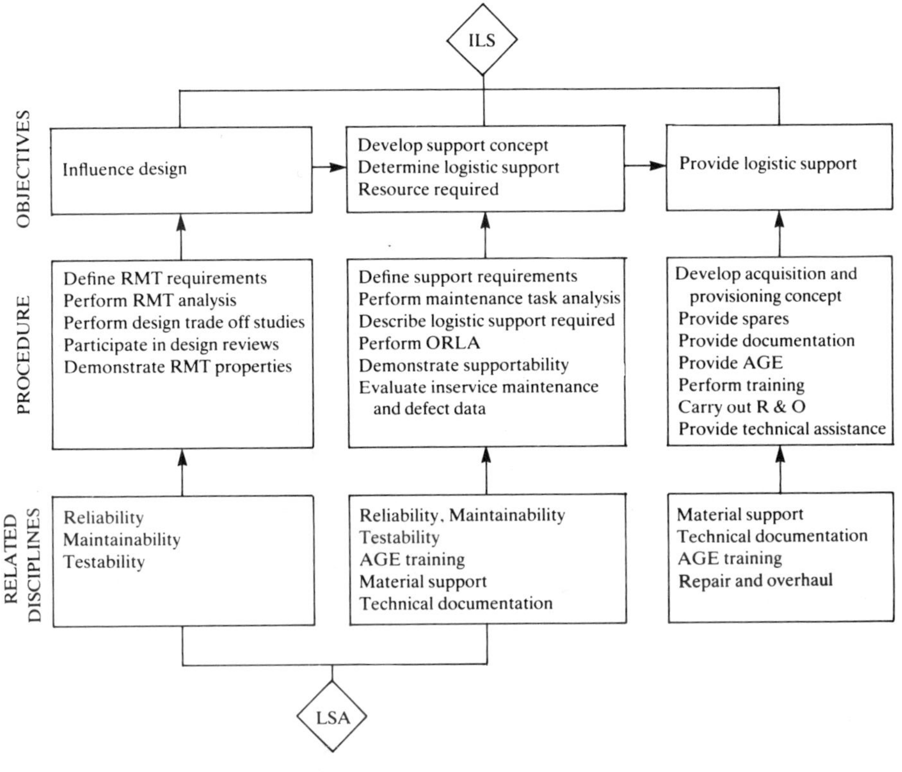

ILS, Integrated logistic support
LSA, Logical support analysis
RMT, Reliability, maintainability, testability
AGE, Aircraft ground support equipment
ORLA, Optimum repair level analysis

Fig. 5.

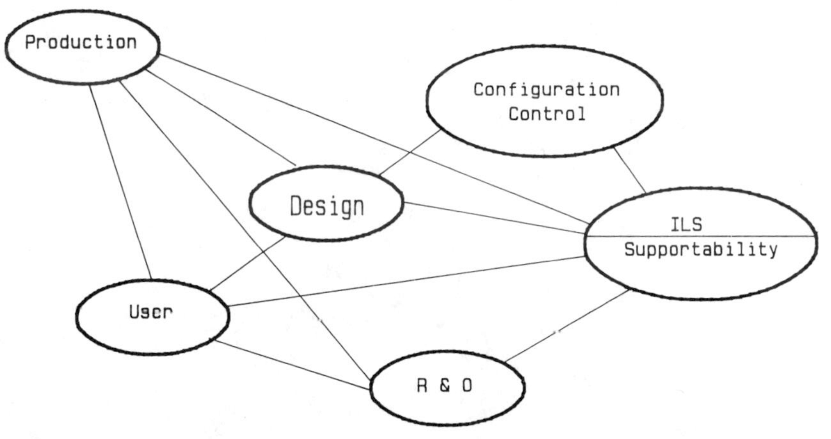

Fig. 6.

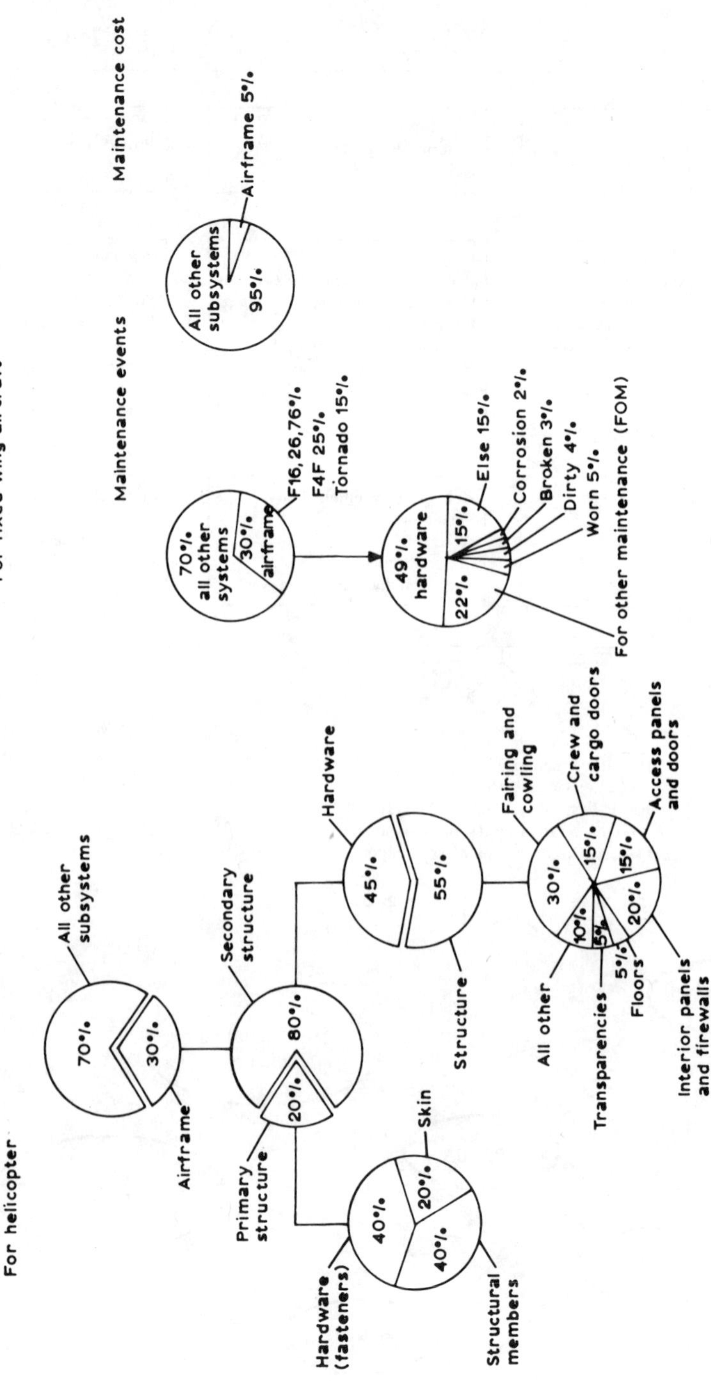

Fig. 7. Airframe contribution to unscheduled maintenance.[1]

TABLE 1[2]

Type of aircraft	*Number of aircraft*	*Total flight hours*	*MTBF*	*Remarks*	*MTBF of a comparable Al-structure*
F15	782	640 000	10 000	Average value of all components	12 000
Speedback			2 260	Range, component dependent	
Stabilator Torque Box			71 430		
			24 000	Average without speedbrake which was considered to be a poor design	
F16					
Horizontal stabilizer		52 752	4 796	Registration period	19 782 Lead. edge flap
Rudder assy		52 752	10 550	May–Oct 1982	5 966 Flaperon assy
Leading edge vert. stab.		52 752	13 188		3 652 Ventral fin
Skins.—vert. stab.		41 630	41 630		
Leading edge vert. stab.		41 630	10 407	Registration period	
Rudder assy		41 630	3 784	Jan. 1981–Nov. 1982	
Skin horiz. stab.		41 630	33 303		
Trailing edge horiz. stab.		41 630	10 407		
Horiz. stab.		41 630	2 250		

TABLE 2

Components	*MTTR*	*Repair level*
Boeing 737 spoiler (Graphite/epoxy reinforced A1)	19	—
Boeing 737 spoiler (A1 design)	20·4	—
Tornado spine hood (CFC/epoxy reinforced honeycomb sandwich structure)	20	Depot
Other tornado components (Glass fibre reinforced honeycomb sandwich).	6–10	Organizational
Structures like flaps, access doors, landing gear door, and glass fibre components	15–70	Depot

derived from experience of previous projects, from basic knowledge or from logical considerations.

Design factors shown in Tables 3–5 and Figs 8–10 are mainly related to the design concept, components, material, method of assembly, maintenance, accessibility, load intensity and interface constraints.

5 DAMAGE CAUSES AND PREVENTIVE MEASURES

Experiences with composite material in aircraft have shown that most of the damage is caused by impact. The majority of impacts result from poor maintenance such as striking the aircraft with ground support equipment and dropping tools. The critical areas of damage are edges, surfaces and access panels. The most effective preventive measures for such damage are to avoid the use of thin-skin material, to protect edges, and to train maintenance personnel how to work with composite material. A list of the major damage causes, critical areas of damage and preventive measures is given in Table 6.

6 DAMAGE TOLERANT DESIGN

A basic requirement of a composite structure is that barely visible damage should not be allowed to reduce structural integrity during service life. To realize this requirement the following problems must be solved. First, what is the expected upper limit of the impact energy during the operation of an aircraft and consequently, what is the extent of the barely visible damage, which has a strong effect on design allowables (i.e. for significant parts of the

TABLE 3
Maintainability Design Factors[3]

Design factor		*Effect on maintainability*
Construction form	Monolithic sheet	Repairability good when both sides of panel exposed. Simple, well-established repair procedures.
	Stiffened sheet	Presence of stiffeners makes repair more complex.
	Sandwich	Bond failures between core and facings may be difficult to detect. Repairability generally good; damaged core can be filled in and patched over. Absence of complex shapes and curvatures simplifies repair.
Stiffener form	Open section	Easiest to repair because all surfaces are exposed.
	Hollow core	Repair limited to external surfaces because of inaccessibility to interior.
	Foam core	May offer slight advantage over hollow core since core material can be filled-in to provide a mould for cure-in-place repair.
Method of assembly	Co-cured	Joint is permanent; must be cut apart for repair.
	Adhesive bond	Absolute cleanliness required to achieve good bond; difficult to implement in field environment. Vertification of integrity of repair difficult under field conditions. Some adhesives require refrigeration and have limited shelf life. High skill required.
	Mechanical fasteners	Easiest type of joint to disassemble. Caution needed in use of mechanical fasteners for repair to avoid introducing stress concentrations and to avoid incompatibility of materials (aluminium and graphite for example).
Contour	Double curvature	Material must be stretched or shrunk to conform to 3-dimensional surfaces; special moulds required. Labour to laminate contoured parts related to amount of curvature.
	Wrapped surface	Less difficult to laminate than double curvature; mould required.
	Flat surface	Easiest to repair; no moulds required.
Accessibility	Restricted	Poor accessibility impedes inspection. Restricted access impedes on-aircraft repairs; limits the use of equipment; increases the probability of faulty repair; adds to repair time.
Load intensity	Lightly loaded	Quality of repair less critical than more heavily loaded structures; visual inspection of repair adequate.
	Moderately loaded	Quality of repair is important; verification of integrity via non-destructive inspection techniques may be necessary.
	Heavily loaded	Quality of repair is critical; usually requires replacement or custom-engineered repair. Verification of integrity via non-destructive inspection techniques will be necessary.
Interface constraints	Equipment mounting provisions and cutouts	Requirements for equipment interchangeability impose dimensional constraints on repair (flush surfaces for example).

TABLE 4
Reliability Design Factors[3]

Design factor		Effect on reliability
Construction form	Monolithic sheet	More flexible than sandwich; greater impact strength.
	Stiffened sheet	Similar to monolithic sheet.
	Sandwich	Facings thinner than equivalently loaded monolithic panels; more easily punctured. Less impact resistant than monolithic sheet. Bond failures may occur between core and facings due to overstress or impact.
Stiffener form	Open section	Less stable than closed section forms; more vulnerable to twisting or buckling type failures.
	Hollow core	Closed section more stable than open section; less vulnerable to buckling or twisting type failures.
	Foam core	Similar to hollow core.
Method of assembly	Co-cured	Excellent bond strength due to resin intermixing.
	Adhesive bond	Simple structural joint. Cleanliness and quality control critical to achieving structural integrity.
	Mechanical fasteners	May loosen and cause fretting or separation of joint.
Contour	Double curvature/ wrapped surface	Sharp exposed radii may be vulnerable to impact.
	Flat surface	Least vulnerable to impact.
Accessibility	Restricted	Inability to inspect properly may allow flaws or damage to progress to advanced stages.
Load intensity	Lightly loaded	Damage has minimal effect on structural integrity; adjacent structure supports load in event of localized damage. Most easily damaged to lightweight construction.
	Moderately loaded	Structural integrity more seriously affected by damage.
	Heavily loaded	Any damage is critical.
Interface constraints	Equipment mounting provisions and cutouts	Local structure reinforcement for equipment adds to complexity; introduces potential failure modes. Affected by loads existing in structure and introduced at interface.

TABLE 5[3]

Design factor	*Potential R & M concern*	*Potential design solution*
Load intensity related		
Heavily loaded tension members (longerons, etc.)	Loss of strength due to moderate impact damage	Provide sufficient strength to withstand normal service damage
Heavily loaded structure	Damage propagation (cracks, delamination)	Use increased damage tolerance materials Use crack arrestors Design for low stress levels
Material compatibility related		
Graphite-to-metal interface	Corrosion	Avoid condition
		Provide insulation at facing surfaces
Interfaces related		
Equipment mounts and supports	Loosening, fretting and fastener hole wear	Reinforce mounting areas
Edges of aluminium honeycomb panels	Moisture entry and corrosion	Provide sealants to prevent moisture accumulation
		Use non-wicking adhesives
		Use corrosion resistant materials
Cutouts	Edge delamination	Increase thickness
		Add protective doublers
		Avoid conditions where thin edges are directly exposed
Maintenance related		
Co-cured and adhesively bonded joints	Disassembly	Provide means for in-place repair
		Use modular design concept (repair strips)

(contd)

TABLE 5—*contd*

Design factor	*Potential R & M concern*	*Potential design solution*
Environment related		
Composites	Environmental deterioration (strength, stiffness)	Apply appropriate strength reduction factors during structural design and qualification
	Fatigue damage	Same as above
	Burns, delamination due to lightning strikes	Provide conductive covering (flame spray, aluminium tape or wire mesh)
Component related		
Fairing, cowling	Edge and corner damage due to rough handling	Increase thickness
		Add protective doublers
Fairings	Abrasion/fretting damage in high vibratory environment	Reinforce wear points (attachment surfaces)
		Add protective doublers
		Increase thickness
Doors	Buckling, warping, misalignment due to slamming	Provide sufficient strength to withstand loading
		Verify with testing
	Hinge-to-structure damage due to slamming open	Same as above
Walkways, floors	Abrasion damage	Provide wear resistant surface coating (polyurethane paint)
		Provide sacrificial or protective covering
		Substitute metal facings
Joints and fittings	Damage due to impact	Provide sufficient margin to withstand normal service abuse

Construction-related		
Sandwich panels	Loss of compression stability (buckling failures) due to light to moderate impact	Increase moment of inertia to reduce sensitivity to impact defects
		Use impact resistant facing material or increase facing thickness
		Add protective coverings
		Use monolithic construction
Aluminium honeycomb panels	Moisture entry around inserts and corrosion	Use non-metallic or corrosion resistant materials
		Add sealant around holes
		Replace inserts with anchor nuts attached to bonded zee closures
Nomex honeycomb panels	Broken/crushed core due to impact	Increase panel thickness
		Increase core density
		Increase facing thickness
		Use monolithic construction
Composite-faced honeycomb panels	Surface fractures and penetrations due to light to moderate impact	Increase facing thickness
		Increase panel flexibility (energy absorption capabilities)
		Add protective covering
Aluminium-faced/aluminium honeycomb panels	Denting due to light to moderate impact	Use rigidized skins
		Increase skin gauge
		Increase core density
		Use unidirectional fibreglass facings
		Use Nomex core
		Add protective covering

(contd)

TABLE 5—*contd*

Design factor	*Potential R & M concern*	*Potential design solution*
Stiffeners, open-section	Twisting, buckling type failures	Provide intermediate supports
		Substitute closed-section members
		Use sandwich construction
Gauge/thickness related		
Thin monolithic composites	Punctures at low energy levels	Increase material thickness
		Use more impact-resistant materials
Thick monolithic composites	Subsurface damage (delamination) due to heavy impact	Provide sacrificial facing
		Provide multiple load paths—redundancy
		Limit stress levels to impede damage propagation
Minimum gauge construction	Damage susceptibility	Avoid composite materials less than 0·020 inch thick
		Alter structural geometry to enable use of thicker materials without compromising structural efficiency
Thin unidirectional graphite/epoxy	Face splintering due to impact	Use woven graphite fabric for outermost ply on exposed surfaces
Fastener/attachment related		
Mechanically fastened composite structure	Fastener overtorque damage	Provide grommets
		Use large head fasteners
		Avoid flush head fasteners
		Provide adequate thickness
	Fastener hole elongation and tearout	Provide adequate reinforcement at holes
		Use large diameter fasteners
		Provide large edge distance
		Insert metal shims in laminate

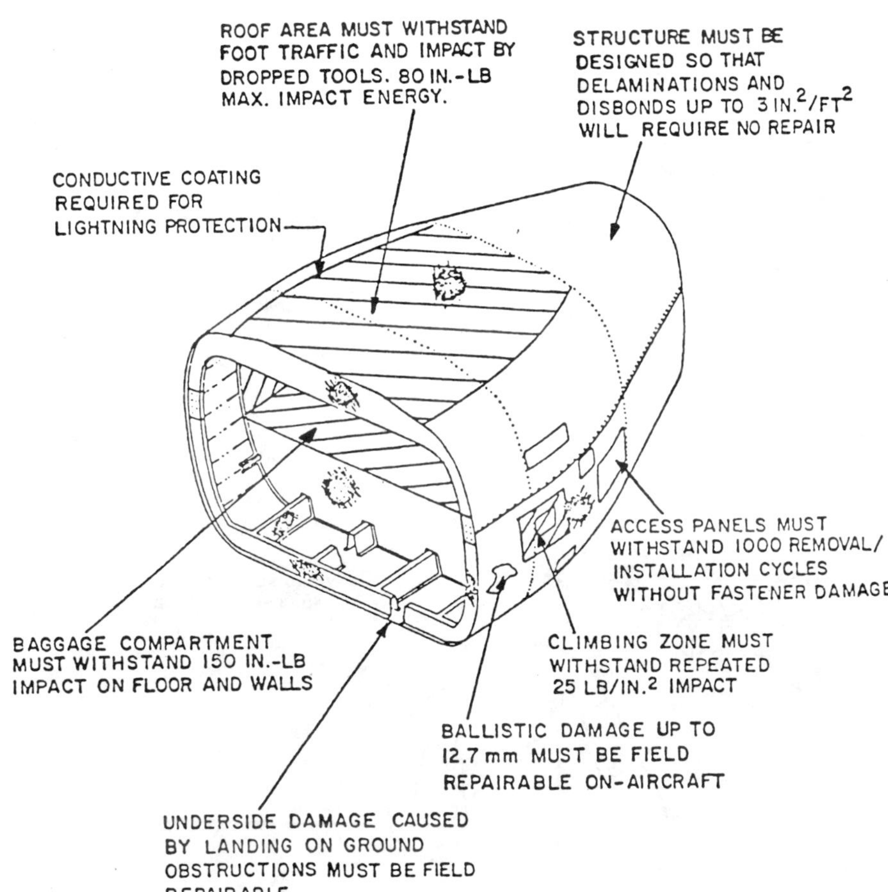
ROOF AREA MUST WITHSTAND FOOT TRAFFIC AND IMPACT BY DROPPED TOOLS. 80 IN.-LB MAX. IMPACT ENERGY.
STRUCTURE MUST BE DESIGNED SO THAT DELAMINATIONS AND DISBONDS UP TO 3 IN.^{2}/FT^2 WILL REQUIRE NO REPAIR
CONDUCTIVE COATING REQUIRED FOR LIGHTNING PROTECTION
ACCESS PANELS MUST WITHSTAND 1000 REMOVAL/INSTALLATION CYCLES WITHOUT FASTENER DAMAGE
BAGGAGE COMPARTMENT MUST WITHSTAND 150 IN.-LB IMPACT ON FLOOR AND WALLS
CLIMBING ZONE MUST WITHSTAND REPEATED 25 LB/IN.^{2} IMPACT
BALLISTIC DAMAGE UP TO 12.7 mm MUST BE FIELD REPAIRABLE ON-AIRCRAFT
UNDERSIDE DAMAGE CAUSED BY LANDING ON GROUND OBSTRUCTIONS MUST BE FIELD REPAIRABLE

Fig. 8.[1]

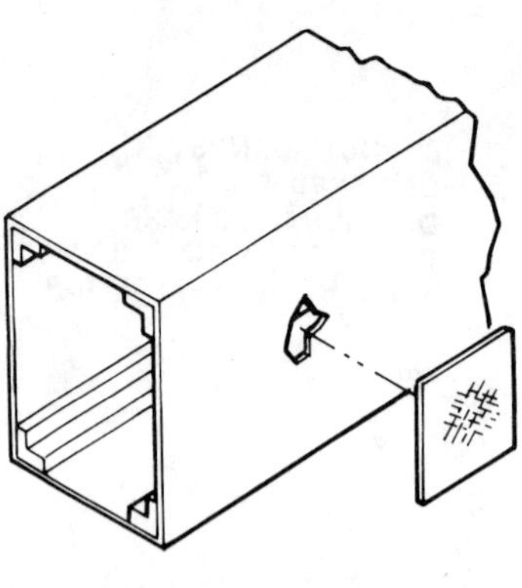
Flat surface easiest to repair

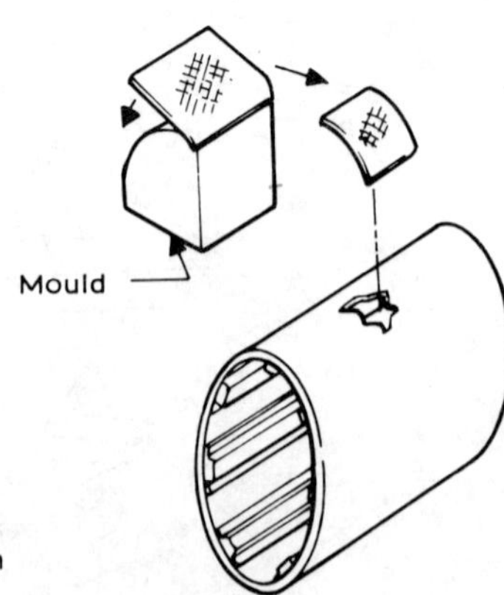
Mould
Wrap surface. Mould
required to laminate patch

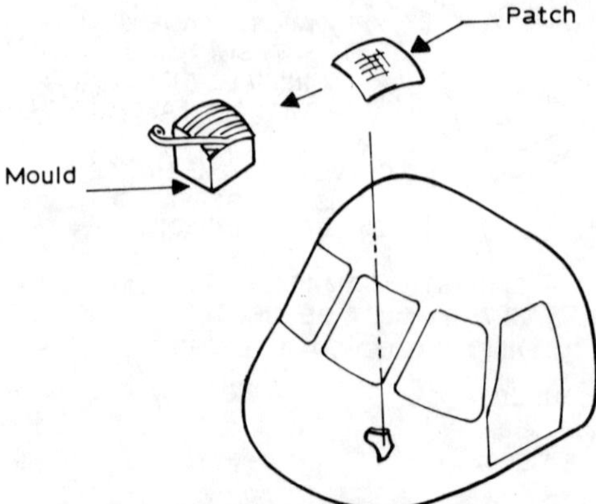
Patch
Mould
Compound curvature. Most difficult repair.
Materials must be laid up in strips

Fig. 9.[1]

MINIMAL EFFECT ON
STRUCTURAL INTEGRITY
LIGHTLY LOADED STRUCTURE
STANDARD
REPAIR
MODERATELY LOADED
STRUCTURE
COMPLEX
CRITICAL
LAYUPS
HEAVILY
LOADED
CONSTRUCTION
HEAVILY LOADED STRUCTURE

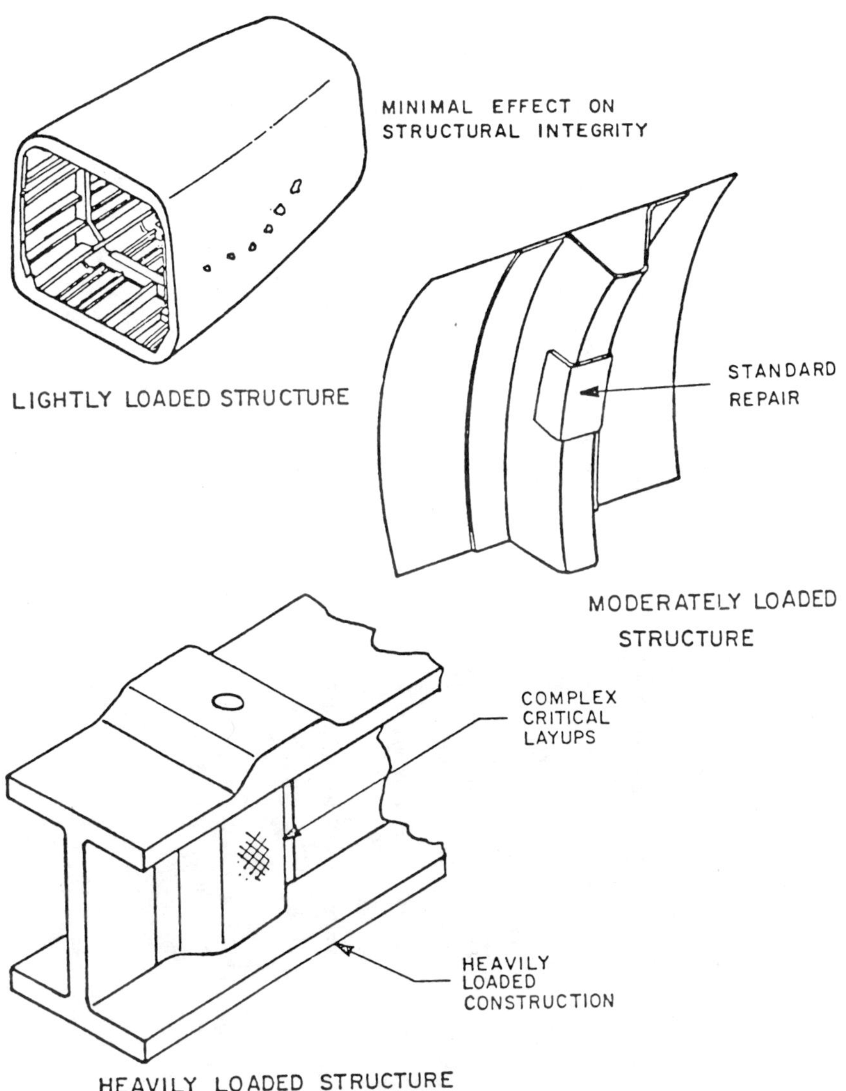

Fig. 10.[1]

TABLE 6

Causes for external type damages	*Critical damage areas*	*Failure modes*	*Preventive measures*
Ground support equipment	Exposed wing edges, areas around cargo and access doors, engine cowls	Delamination, penetration	Preferable use of monolithic instead of thin skin sandwich structures. GSE design shall account for low impact damage.
Dropping tools	Upper wing surfaces, upper fuselage	Delamination	Tolerance of composite structures. Training of maintenance personnel.
Debris from runway	Inboard flaps and inboard lower wing and fuselage surfaces	Delamination	Avoidance of thin skin sandwich structures.
Bird strikes	Leading edges, frontal areas	Delamination, penetration	Leading edges in metal design or at least metal protected.
Hailstones	Leading edges, frontal areas, upper fuselage and wing surfaces	Delamination puncture	Avoidance of thin skin sandwich structures. Metal construction or metal protection of leading edges and frontal areas.
Rain erosion	Unshielded edges of panels and doors, leading edges and front fuselage	Erosion of surface material	Leading edge protection by metal or abrasion resistant coating, e.g. rubber.
Lightning strike	Mostly leading and trailing edges	Delamination, resin burns	Applying cocured metal mesh or metal strips for lightning protection.

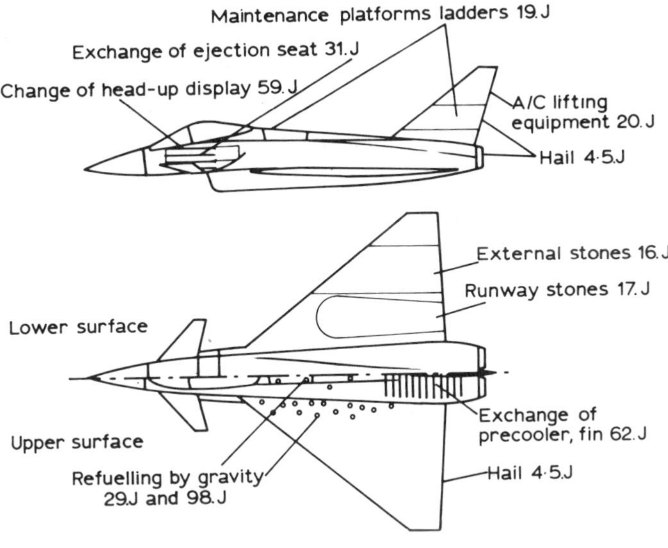

Fig. 11.

composite structure the design strain will be determined by barely visible impact damage).

One should also ascertain a proper definition of 'barely visible' damage. A study was carried out at Messerschmitt-Bölkow Blohm (MBB) analysing the environment with respect to impact energy which may affect surfaces during maintenance or operation of the aircraft (Fig. 11). From this study a representative impact energy was derived for impact tests. Design allowables have to be defined such that the residual strength of impacted components is not below the design ultimate. It is evident that the impact energy level chosen for the project will influence structural weight and supportability.

7 REPAIRABILITY REQUIREMENTS

Repairability and detectability are important features of maintainability and therefore for supportability. Several repair methods have been developed. The assessment of these methods has lead to encouraging results, i.e. most of the repaired components have achieved the required strength properties. Besides strength recovery other repair requirements listed below are also taken into consideration.

—100% Design strength recovery with minimum change in stiffness.
—Serviceability for the aircraft design temperatures and environment.

—Minimum weight penalty.
—Minimum aerodynamic contour changes for surface repairs.

Most of the repair programmes performed at MBB or presented by the literature have been performed under laboratory conditions. In the longer term it will become a hard customer requirement to develop composite components and repair methods such that maintenance, especially repair on and off aircraft, can be performed by the user. To achieve this target, specific requirements, which are listed below must be met.

Field repair level

—standardized, prefabricated external patches
—RT or low temperature cure
—only vacuum
—external patches only
—standardize repair kits and manuals
—only a very limited number of materials should be used
—long shelf life of materials is required

Depot and factory repair level

—Engineering actions are necessary
—Develop repair materials for high performance structural repairs
  —vacuum cure must provide low porosity bondlines and laminates
  —sensibility to moisture must be low
  —low temperature cure should be possible
  —reduce number of materials
  —use one set of repair design allowables
—Repair material must not be identical with parent material
—Develop similar repair procedures (scarf angles, stepped joints, type of external patches, cleaning of repair sites, redrying, prefabrication of repair patches, bolted repairs, etc.).

Use qualified materials and procedures for all aircraft; avoid qualification and certification of repairs for every new aircraft.

REFERENCES

1. IDA/OSD, Reliability and Maintainability Study Vol. IV, Steering Group Report. Institute for Def. Analyses, Alexandria, VA, Nov. 1983.
2. Structural Composite Working Group Report, IDA Record Doc. D31, 1983.
3. Cook, Th. N. & Kay, B. F., Advanced Structures Concepts R & M/Cost Assessments USA RTL-TR-79-18, Sept. 1979.

Composite Structures **10** (1988) 37–50

Supportability of Composite Airframes: Civilian and Military Aspects

Stefan Kupczyk

Dornier GmbH, Dept. SK51, PO Box 1420, D-7990 Friedrichshafen, FRG

ABSTRACT

The advantages of composite materials are, weight saving, good stiffness behaviour, excellent surface quality and the capability of cost-efficient production. Yet, apart from the advantages of these advanced materials, their use also involves risks and problems. Of major concern are operating problems such as reliability, maintainability, repairability, availability and also producibility of aircraft components made of composite materials. These problems are described in the following chapters and ways to overcome these problems are shown.

1 COMPOSITE COMPONENTS IN DORNIER AIRCRAFT CONSTRUCTION

Within the frame of civil and military programmes Dornier has gained extensive know-how and expertise in the use of composite materials for aircraft production. The CFRP air brake (Fig. 1) for Alpha Jet weapon system (Fig. 2) has been flown and mass-produced for several years. Under the terms of a flight test programme, a carbon fibre-reinforced plastic (CFRP) horizontal tail (Fig. 3) is being investigated. A CFRP rudder (Fig. 4) has been undergoing endurance testing for several years and the development of a CFRP wing was started nine years ago (Figs 5 and 6).

In civil aircraft programmes, composite technology has also been favoured. Numerous components of the Dornier 228, a utility/commuter aircraft, are made of composite materials, primarily for secondary structure applications (Fig. 7). The aileron (Fig. 8) is a fitting example with its leading

Composite Structures 0263-8223/88/$03·50

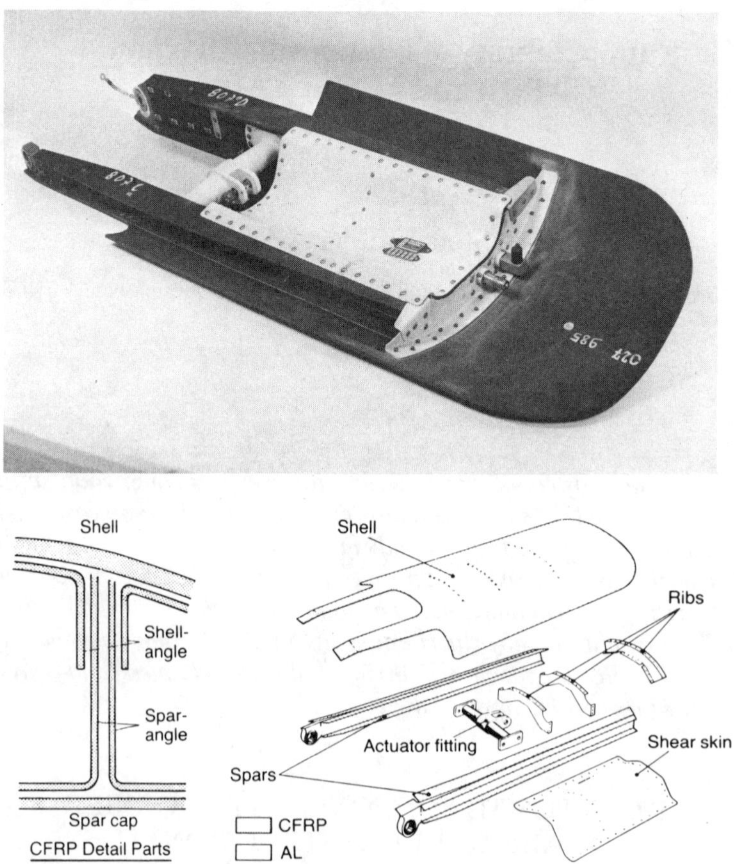

Fig. 1. Alpha jet CFRP speed brake.

edge of CFRP and the spar box of synthetic fibre-reinforced plastics (SFRP). The spar, the apex strips and the load transfer fittings are made of aluminium alloys.

With the follow-on model, the Dornier 328 (Fig. 9), the use of CFRP material for the tail cone structure and the tail unit is projected. Sections of the wing, various transitions and the fairing will be made from a GFRP sandwich of SFRP. A CFRP/Kevlar hybrid will be used for the control surfaces. Under the terms of a pre-programme a CFRP fuselage section (Fig. 10) has been built which is currently undergoing dynamic internal pressure tests. This test is especially designed to reveal the behaviour of production defects, impact damage and the behaviour of repairs when subjected to pulsating internal pressure loads. Under subcontract to Airbus

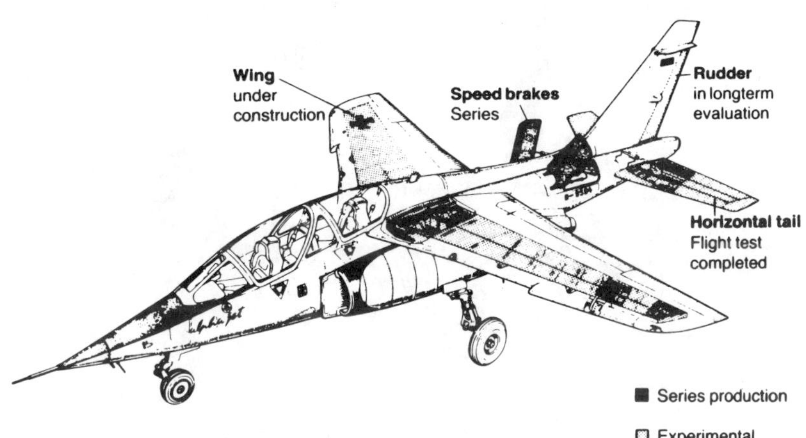

Fig. 2. CFRP components of the Alpha jet.

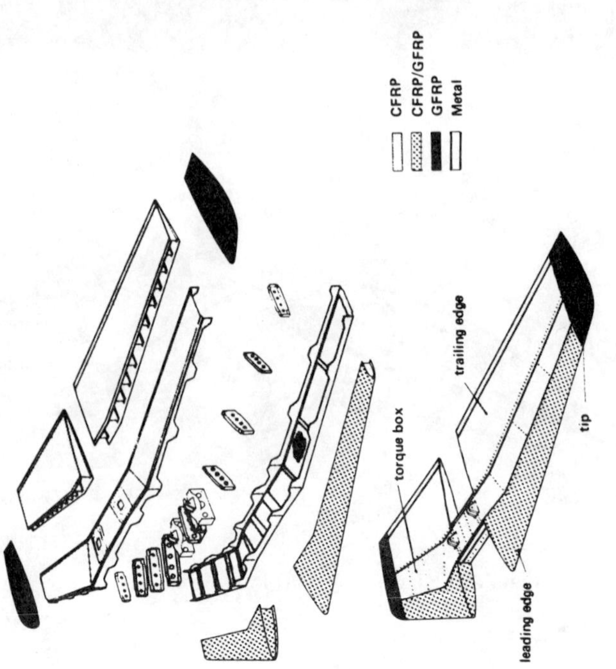

Fig. 3. Composite horizontal tail unit for the Alpha jet.

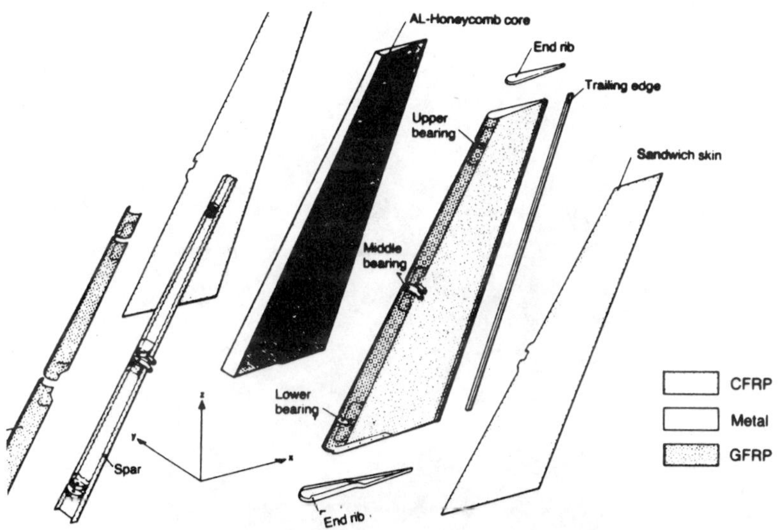

Fig. 4. CFRP rudder for the Alpha jet.

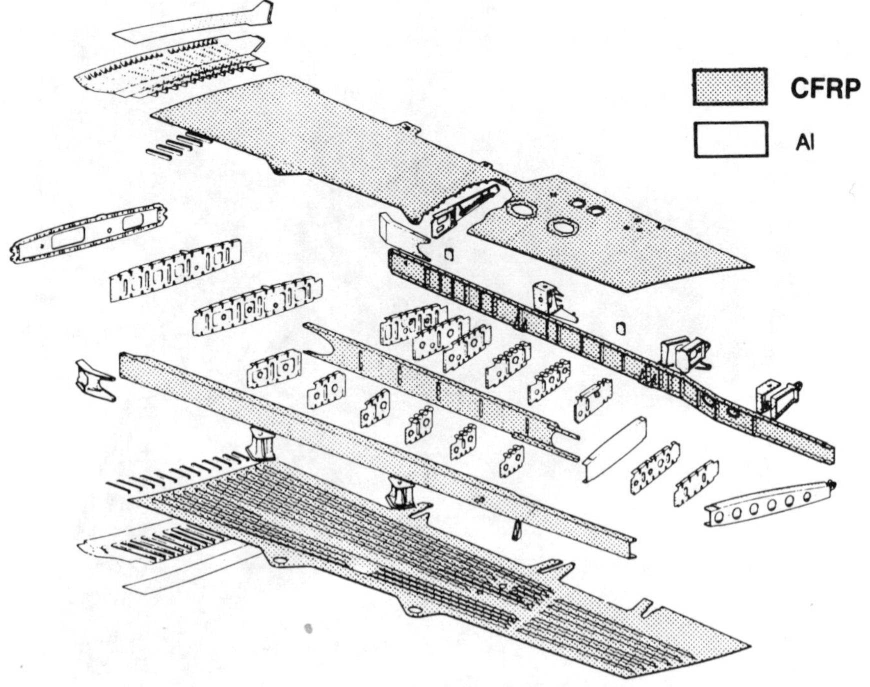

Fig. 5. CFRP wing of the Alpha jet (experimental).

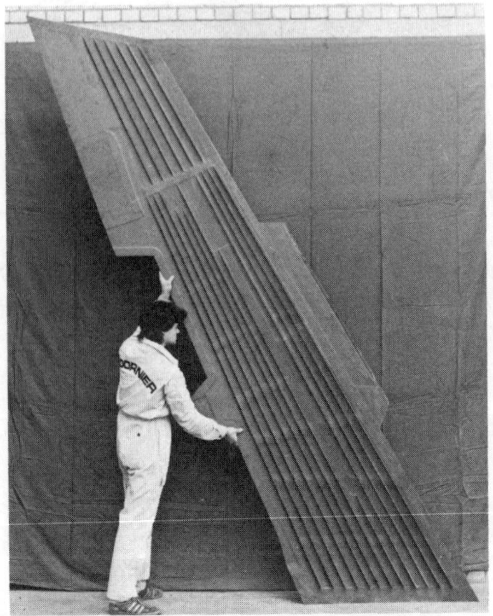

Fig. 6. CFRP wing-panel of the Alpha jet.

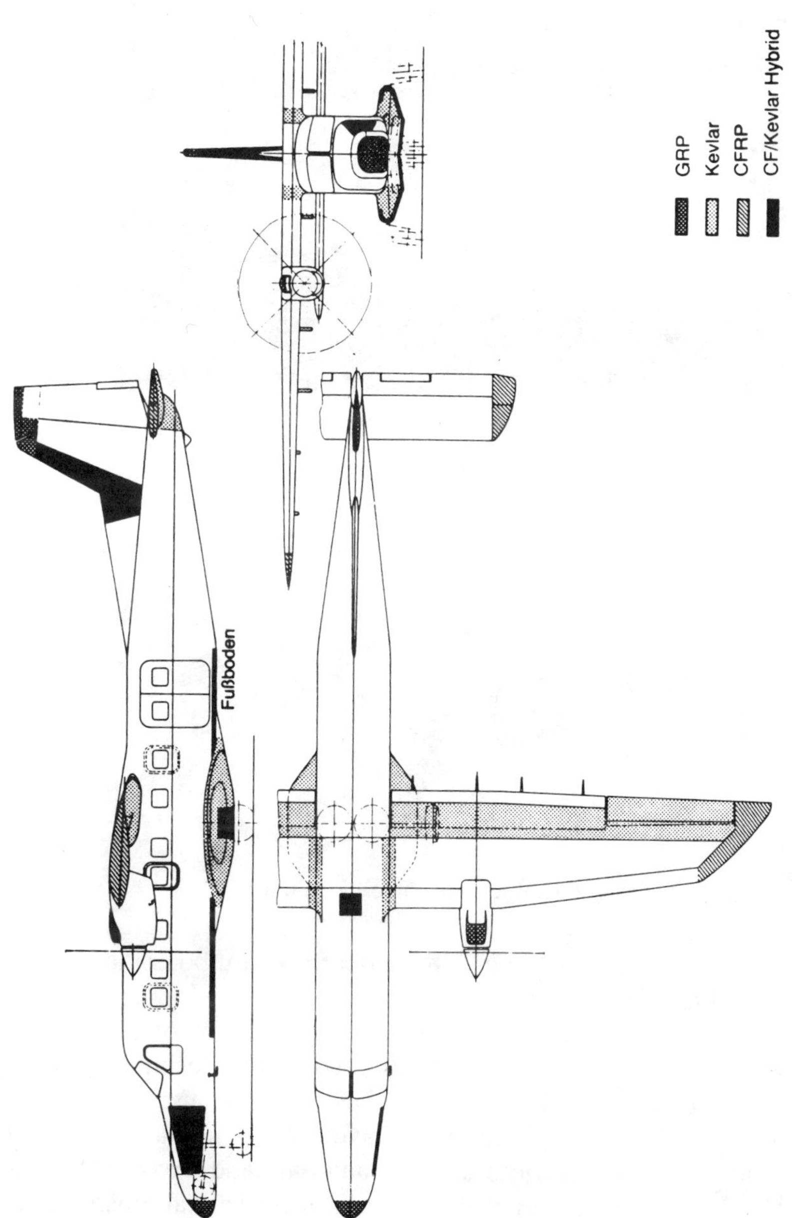

Fig. 7. Composite components of the Dornier 228.

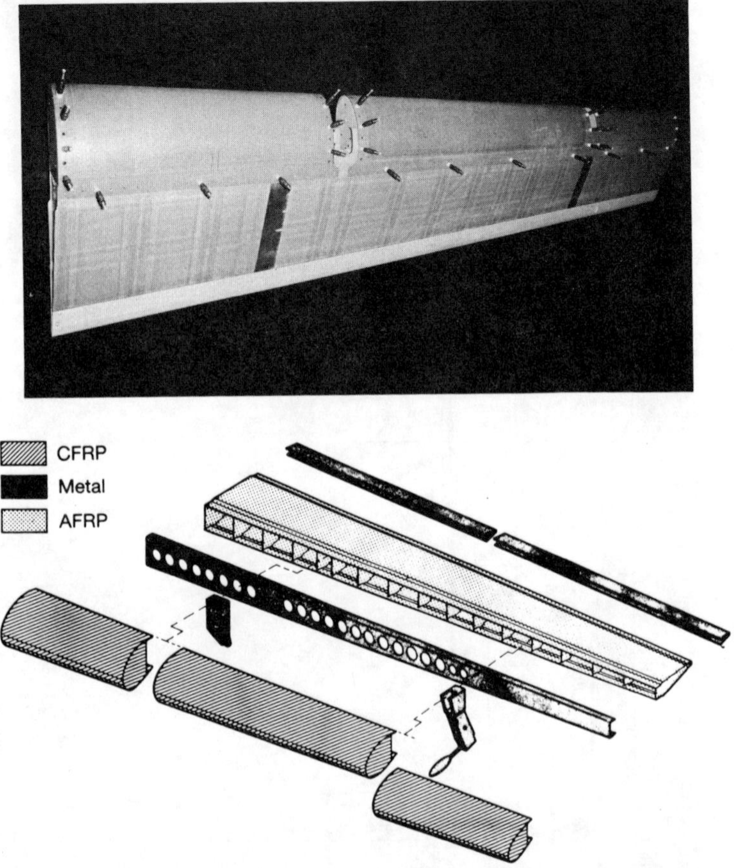

Fig. 8. Dornier 228 composite aileron.

Industries, Dornier is building the CFRP inboard and outboard flaps for the Airbus A320 (Fig. 11).

This brief summary is designed to provide an overview of a part of the know-how offered by Dornier in the field of composite technology. Many advantages exist in using advanced materials, but their use also involves risks. The advantages include modern, favourable construction, weight saving, good stiffness behaviour of the materials and excellent surface quality. They could not be achieved with conventional metal constructions. But the risks involved pose tremendous problems. A major part of the issues concerns operating problems such as reliability, maintainability, repairability, availability, and also producibility of aircraft components and composite materials.

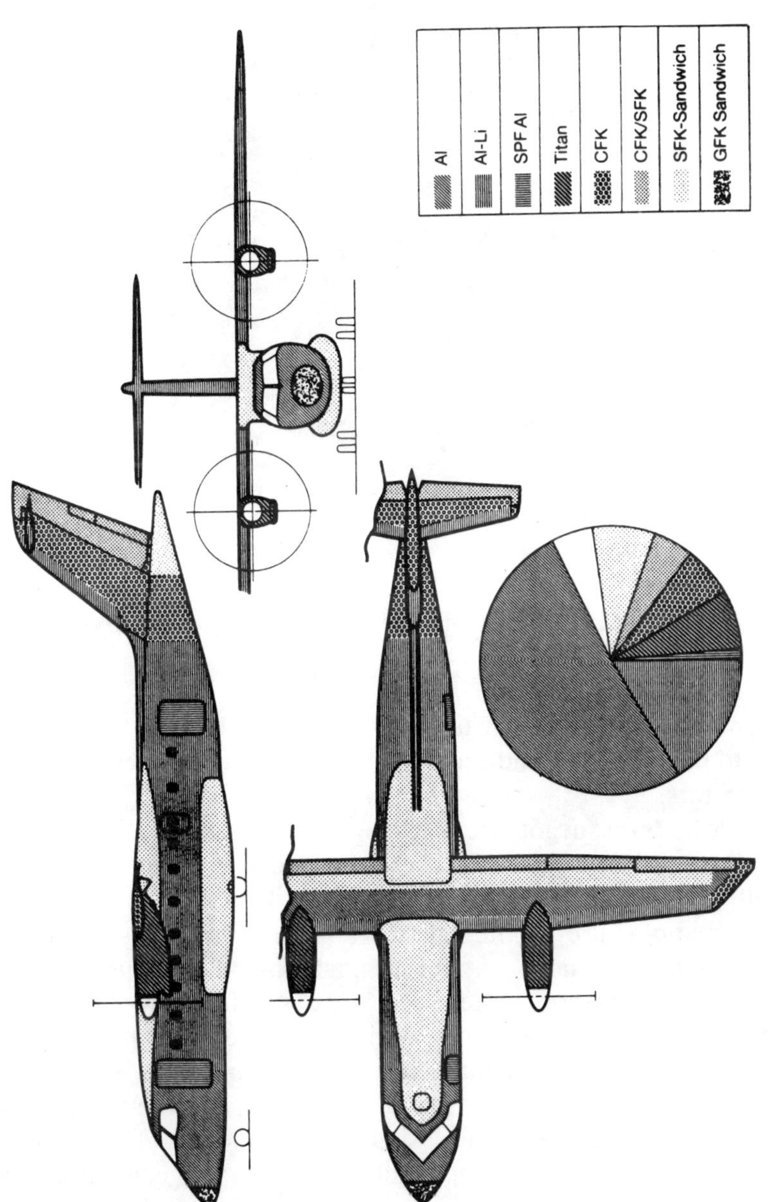

Fig. 9. Dornier 328—advanced materials and technologies.

Fig. 10. CFRP fuselage test segment.

2 OPERATING PROBLEMS WITH AIRFRAMES MADE FROM COMPOSITES

The problems are twofold: firstly the high cost involved in the manufacture of composite components, and secondly the operating problems faced by aircraft operators.

Through long-term negotiations with suppliers of CFRP prepregs, material prices for the Do 328 project could be lowered by 50%. Additionally, due to intensive development efforts in process engineering towards the manufacture of integral CFRP components, it is possible to achieve the same component cost as for integral metal constructions. This demands a quality assurance system which is cost-effectively embodied in the entire production process. To avoid duplication of effort the outgoing inspection by the material's manufacturer is combined with the Dornier incoming inspection. Automated non-destructive testing methods further reduce quality assurance cost. The share of cost for quality assurance measures could thus be cut from an initial 100% to 14% of component part costs. The advantages achieved are evident. In the case of the Do 328 tail cone structure, which is made from CFRP, over 30% of weight is saved compared with metal structural components, and the manufacturing costs are almost the same.

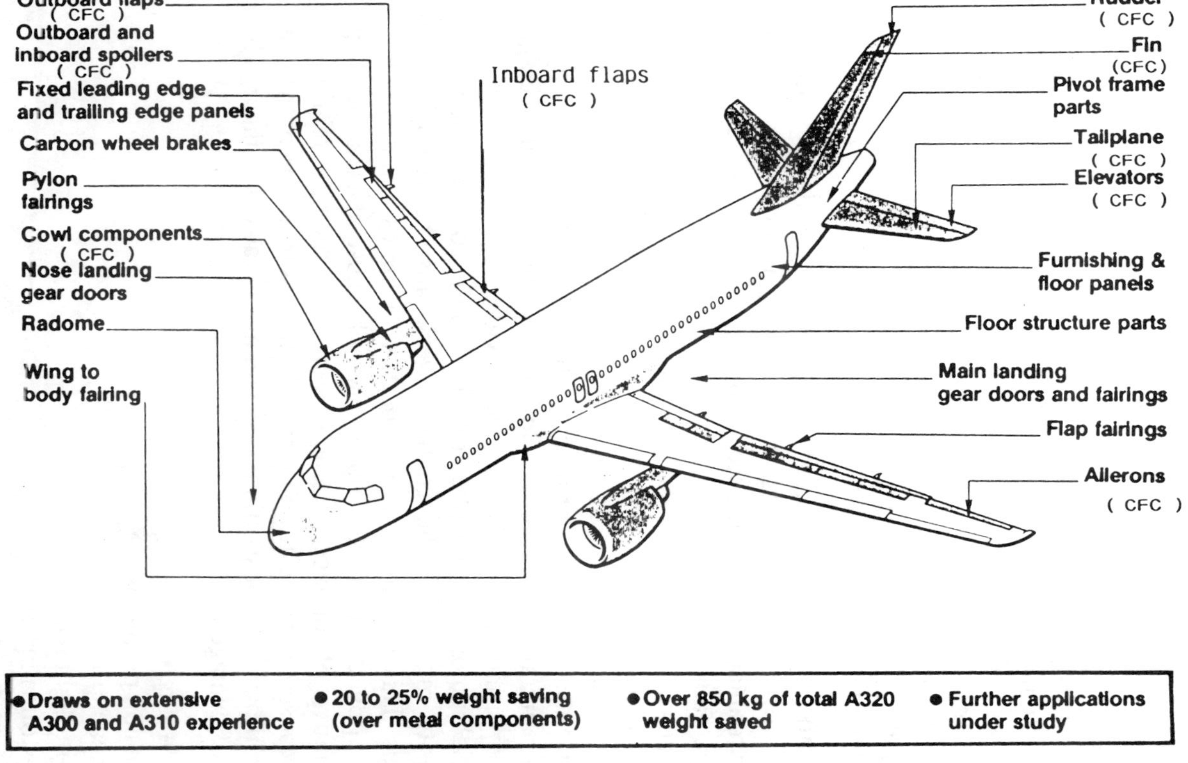

Fig. 11. Airbus A320—wider application of composite materials.

It is also important to consider the operating problems encountered with composite components. The following issues must be addressed in this context:

—handling
—corrosion problems
—coating possibilities
—inspection
—maintenance
—detectability of damage
—damage assessment
—damage behaviour
—repair
—availability of spares

Aircraft operators complain that the majority of damage occurs during service work in between flights and during maintenance activities. The primary damage cause is the frequent contact of the aircraft with baggage vehicles, gangways and work platforms. Exposed areas include passenger and cargo compartment doors. With large commercial airliners the baggage compartment doors usually have so many dents after five years that they must be replaced. Frequent damage is caused by aircraft mechanics who inadvertently drop tools during inspection, maintenance and repair work. With an aircraft the size of the Do 328 this amounts to an average of one damage per month caused by service activities.

Special importance must be attributed to the coating systems applied to composite materials. The coating layers must be removed at regular intervals. This is performed for inspection purposes, but new coatings may also be applied to maintain appearance or because of a change of operator. The removal of coatings can lead to damage on the fibre-reinforced plastic (FRP) caused by paint stripping or mechanical separation of the paint.

By comparison, substantially less damage is caused in flight. The major causes of damage are

—lightning
—erosion problems
—impact problems
  —in-flight hail
  —stone impact
  —tyre disintegration
  —turbine blade failure
  —battle damage problems

The final cause of damage in the above list demonstrates the difference between military and civilian use.

3 DEFINITION OF THE RANKING OF THESE PROBLEMS

General safety criteria in conjunction with operating problems have absolute priority with military and civil operators. The biggest problem with composite materials is the poor detectability of damage. Delaminations and cracks in the fibres are externally visible only at a very late state when damage is already extensive. This condition, combined with insufficient experience in the long-term behaviour of these types of damage, is a prime reason for the rejection of composite technology by many aircraft operators.

A second item influencing the decision-making process in favour of or against composite components is the grounding times necessary for inspection, maintenance and repair. Including fuel saving and weight reduction, civil users assess these grounding times only in view of their direct operating costs. A civil customer will not incorporate advanced technologies if they do not offer lower DOC. The maintenance, inspection and repair facilities available to the civil operator also play a crucial role in this context. Major airlines, too, do not accept frequent inspections using Non-Destructive Testing (NDT) methods. Likewise, repairs requiring an autoclave are not accepted.

With respect to these problems military users place priority on the fulfilment of mission requirements. If the survivability of airborne systems is improved through the use of composite components, more cost-intensive inspection methods, maintenance or repair methods are accepted. This, however, requires the development of methods and processes which allow inspection and repair also on airfields.

The solutions to the problems discussed in the course of this workshop will decide on the increasing use of composite technologies in the years to come. Thus, the issue of operating problems must be ranked very high.

4 IDENTIFICATION OF GUIDELINES OR SOLUTIONS TO THE PROBLEMS

The opportunities provided by the introduction of composite materials demand great efforts on the part of manufacturers and operators. The aircraft makers, however, must make the first step, concentrating efforts on the design and construction of composite components. Small design stresses or design strains can prevent the potential extension of damage.

A better approach, however, would be to tackle the causes of damage. Stone deflectors on the landing gear, for instance, should be mentioned in this context. Service vehicles such as platforms and ladders could also be provided with appropriate protection systems to prevent damage during service work. Particular effort must be made in the development of maintenance and inspection methods. Attention must be paid to executing such methods reliably, cost-effectively, in a timesaving manner and by simple means.

Another important subject to focus on is the provision of repair methods. They must comply with the following demands:

—execution by simple means, e.g.
  —infrared heater
  —riveted sheet metal or parts made from composite materials
  —cold setting adhesion systems
—execution by personnel without special qualification
—little expense with regard to time and cost
—high static and dynamic strength

Only when such repair methods are available will the customers opt for an increasing use of composite materials. Not only are the efforts of aircraft makers called for, but operators should contribute their share. In future, they must improve on their training of service and maintenance personnel to be able to profit from the benefits offered by advanced technologies.

5 PROSPECTS

Particularly for aircraft construction, composite technologies offer an opportunity to produce components that are lighter, stiffer, aerodynamically superior, more resistant to corrosion and less critical with respect to fatigue. This technology, however, still involves risks which can be technically overcome in the majority of cases but which are still met with distrust from many sides. One reason for this is undoubtedly insufficient long-term experience and the fear of high operating cost due to potential operating problems.

If increasing use is to be made of the impressive advantages offered by composite technology, efforts must be concentrated on finding a solution to these problems. Approaches include:

—damage-tolerant structures
—simple, reliable inspection methods
—appropriate maintenance concepts and
—simple, timesaving and cost-effective repair methods.

Composite Structures **10** (1988) 51–73

Assessment of the Effect of Impact Damage in Composites: Some Problems and Answers

R. Jones, J. Paul

Department of Defence, DSTO, Aeronautical Research Laboratories, PO Box 4331, Melbourne, Victoria, Australia

T. E. Tay & J. F. Williams

Department of Mechanical and Industrial Engineering, University of Melbourne, Victoria, Australia

ABSTRACT

In order to provide through-life support for structural components, fabricated from advanced composite materials, it is desirable to establish a damage tolerance methodology. The availability of this methodology would greatly aid the design of safe, efficient composite structures as well as their management in service. There are, however, many difficulties in meeting this objective, including the multiplicity of failure modes in the composites, the numerous types of potentially significant defects which may arise during manufacture or in service, and the sensitivity to moisture and temperature. This paper discusses the analysis and testing of impact damage composite laminates. Selected experimental studies in the literature are briefly discussed.

It is shown that uniaxial S–N curves for damaged laminates have a generic shape with a pronounced threshold level. Indeed, it is clear that further research is needed to understand the physical reasons for such a threshold value.

1 INTRODUCTION

Graphite–epoxy composites have many advantages for use as aircraft structural materials, including their formability, high specific strength and stiffness, resistance to cracking by fatigue loading and their immunity to corrosion. Thus, besides producing lighter, more efficient structures, the use of composites should provide more durable structures compared with those manufactured from conventional aluminium alloys. Australian interests in graphite–epoxy composite aircraft structures are focused on through-life support of the McDonnell–Douglas F/A-18 for the Royal Australian Air Force. Indeed, a detailed review of damage tolerance from the Australian point of view is given in Ref. 1.

Graphite–epoxy composites, while having the above advantages, are prone to a fairly wide range of defects and damage which may significantly reduce residual strength.[2–6] Of the various types of defects, delaminations (i.e. single or multiple internal cracks whose planes are parallel to the surface of a component)[7–9] arising in service are probably the most insidious because they can cause reductions in compressive strength (up to 65% of undamaged strength[10,11]) and are difficult to detect. Delaminations may develop during service due to the presence of excessive interlaminar shear stresses or through-the-thickness tensile stresses at holes, free edges, in the region of section changes and bonded joints. However, the most important source is impact. Such damage can occur from dropped tools or from stones thrown up from the runway. While impact can cause a significant amount of delamination, often the only external indication is a very slight surface indentation.[12,13] This type of damage is frequently referred to as 'barely visible impact damage' (BVID). The problem of BVID is of particular concern because the damage is unlikely to be discovered unless the region is subjected to non-destructive inspection (NDI). Generally, this employs ultrasonic procedures when BVID is usually readily detectable. However, unfortunately, most routine NDI is likely to be confined to potential hot spots such as critical joints. Frequent full-scale NDI is costly and time-consuming.

It is highly desirable that procedures are available so that the possible occurrence of delamination-type defects is allowed for in the design and certification of composite aircraft structures and in the development of approaches for through-life support, to provide a rationale for setting inspection intervals, particularly for highly stressed regions and also to provide repair/reject criteria.[14,15]

This paper discusses the analysis and testing of impact damaged graphite–epoxy laminates, and their ability to provide information to establish a rationale for inspection intervals.

2 CHARACTERISATION OF DELAMINATION RESISTANCE

Delamination in composite laminates is often a mixed-mode fracture process since in the laminates, both interlaminar tension and shear stresses are present at the delamination front. The interlaminar tensile stresses give rise to a Mode I component (with the associated energy released rate G_I), and the interlaminar shear stresses give rise to a Mode II component (with energy release rate G_{II}). When using the energy release rate approach to characterise delamination, it must be remembered that the growth will be both mixed-mode and non-self-similar.

In attempting to formulate suitable failure criteria based on the energy release rate method, many researchers have designed a variety of experiments for laminated plates with the intention of isolating and characterising pure Mode I, pure Mode II and mixed-mode failure.

2.1 Mode I characterisation

There has been considerable success in characterising Mode I delaminations. The methods most popularly used are the Double Cantilever Beam (DCB) test and the free-edge delamination tensile (EDT) test. The results of these tests on comparable material systems by various researchers are briefly summarised in Table 1a. It can be seen that the values of G_{Ic} are fairly consistent within the type of test method employed, varying between 80 and 240 J/m^2, depending on the matrix material and the nature of the fibre–matrix interface. In interpreting the values of G_{Ic} from DCB specimens, attention should be given to the surface morphology of the delamination area, since extensive fibre bridging[16] or intraply cracking[17] can influence the value of the energy release rate obtained so that it is higher than the value that would have been obtained if the delamination was entirely confined within the matrix-rich region and the plane of initial delamination.

Table 1b briefly describes some of the experimental and data analysis techniques used by researchers in estimating the value of G_{Ic}. For the DCB tests, three methods of analysis are commonly used: they are an analysis based on beam theory,[16,18–21] the area under load displacement curve method[22] and a semi-empirical compliance method.[17] All three methods give reasonable results. The edge delamination tensile specimen has received quite a lot of attention, notably from O'Brien.[23,24] Laminated plate theory and the simple rule of mixtures to compute the stiffness loss due to delaminations have been used to obtain an expression relating the critical energy release rate G_c to the critical strain ϵ_c measured at the onset of delamination. The values of G_c so obtained are often only slightly higher than the value obtained through DCB specimens.[22] This is not surprising since, in EDT specimens, the delaminations initiate at the free-edge where high peel stresses are present, essentially resulting in predominantly Mode I fracture. A disadvantage of EDT-type tests is that they often produce intraply cracking as well as delamination, resulting in complex fracture surface morphologies rather than the 'clean' delamination surfaces encoun-

TABLE 1a
Comparison of Experimental Values of G_{Ic} from the Literature

Material	*Reference*	*Test method*	G_{Ic} (J/m^2)
T300/5208	Wilkins *et al.*[18]	DCB	87·6
T300/5208	Ramkumar & Whitcomb[19]	DCB	102·6
T300/5208	Chai[17]	DCB	86 ± 8
T300/5208	O'Brien[23]	EDT	137
T300/5208	Byers[28]	Modified width-tapered DCB	160–241
GY70/5208	Russell & Street[20]	DCB	83
AS1/3501–6	Russell & Street[20,27]	DCB	110
AS1/3501–6	Jurf & Pipes[29]	Modified Arcan Test	79
AS4/3501–6	Aliyu & Daniel[21]	DCB	198–232
AS1/3502	Bradley & Cohen[16]	DCB	155
AS1/3502	Whitney & Browning[22]	DCB	140
AS1/3502	Nicholls & Gallagher[30]	DCB	105–175
AS1/3502	Bradley & Cohen[16]	CT (Compact Tension)	225
AS1/3502	Whitney & Browning[22]	90° CN	154
AS1/3502	Whitney & Browning[22]	EDT	267
AS4/3502	Bradley & Cohen[16]	DCB	225 extensive fibre bridging observed
AS4/3502	Bradley & Cohen[16]	CT	120
AS4/3502	Whitney & Knight[25]	EDT 50·8 mm wide specimen	103
AS4/3502	Whitney & Knight[25]	EDT 25·4 mm wide specimen	238
AS4/3502	Whitney & Knight[25]	EDT modified with inserts	81–96
AS4/3502	Whitney & Knight[25]	DCB	158
T300/1034C	Donaldson[26]	90° CN	779
T300/934	O'Brien[24]	EDT	216
T300/934	Wang *et al.*[31]	DEN (Double Edge Notched Coupon)	157

tered in undirectional DCB tests. An attempt to modify the EDT specimen by introducing starter delaminations in the form of inserts has resulted in considerable lowering of the value of G_{Ic}.[25] Further, it has also been shown that G_{Ic} obtained through EDT specimens is sensitive to differences in specimen widths[25] (see Table 1a).

The 90° Centre Notch (CN) specimen has also been used by a number of researchers.[22,26] This method relies on the computation of the stress intensity factor K_{Ic} which is, in turn, related to G_{Ic} through an expression involving material constants. Although the experimental and analytical procedures of the 90° CN test are very different from those of the EDT or DCB tests, the values of G_{Ic} obtained through these methods agree reasonably well, perhaps indicating that at least for Mode I behaviour, the test methodologies are acceptable.

TABLE 1b
Description of Analysis and Experiment for Determining G_{Ic}

Reference	*Test method*	*Description*
Wilkins *et al.*[18]	DCB	Compliance method of test procedure employed, i.e. $G = P^2(dC/da)/2W$. Load–displacement curve monitored for static loading. Relationship between compliance and crack length was obtained, $C = 2a^3/3EI = A_1 a^3$. Also, the relationship, $P_c = (G_c WEI)^{1/2}/a = A_2/a$ was obtained. The value of G_c can then be calculated. Both experimental curves of C vs a and P_c vs a were obtained via a least-squares-fit routine to obtain A_1 and A_2.
Ramkumar & Whitcomb[19]	DCB	Similar data analysis as in Wilkins *et al.*[18]
Russell & Street[20]	DCB	Similar data analysis as in Wilkins *et al.*[18]
Bradley & Cohen[16]	DCB	Analysis based on linear beam theory. $G_1 = 8P_s^2/L_c^2/WEI$, where P_s is the applied load, L_c is the length of the split laminate. The analysis was essentially similar to Wilkins *et al.*[18]
Aliyu & Daniel[21]	DCB	Beam analysis method used. The analysis was extended to include transverse shear deformation and kinetic energy effects on the energy release rate. A semi-empirical method based on the assumption of partial elastic foundation support beyond the crack tip was proposed. Effect of crack velocity on energy release rate studied. G_{Ic} found to increase with crack velocity.
Whitney & Browning[22]	DCB	G_{Ic} obtained through the relationship $G_{Ic} = 1/26\Delta c(P_1\delta_2 - P_2\delta_1)$, where P_1 and P_2 are the loads before and after crack extension, δ_1 and δ_2 the deflections before and after crack extension, and Δc the increment in crack extension. The analysis was based on area between the loading and unloading curves.
Chai[17]	DCB	Semi-empirical method of analysis used. An empirical relationship between compliance C and crack length l, $C = al^n$, where α, n are constants, was used to obtain $$G_1 = \frac{n}{2b}\frac{P\delta}{l}.$$

TABLE 1b—*contd.*
Description of Analysis and Experiment for Determining G_{Ic}

Reference	*Test method*	*Description*
Jurf & Pipes[29]	Modified Arean test on Single Edge Notch	The aluminium fixtures enabled mixed-mode loading of the specimen by varying the angle of loading α. For pure Mode I loading, $\alpha = 90°$ and the applied stress normal to the crack plan was $\sigma = P/A$, where P is the applied load and A is the area of specimen on which the load is applied. There was some scatter in the values of K_{Ic} obtained. G_{Ic} was obtained from K_{Ic}.
Byers[28]	Specimens DCB	Width-tapered specimens for comparison of interlaminar fracture toughness of composites with different resin systems. However, the aluminium tabs were actually half-inch-thick plates adhesively bonded to the entire top and bottom surfaces of the laminated specimen. The values of G_{Ic} obtained were much higher than those obtained by others using normal DCB specimens, probably due to increased stiffness introduced by the aluminium plates. The expression $G_{Ic} = 12P^2/(Eh)(a/b)^2$ was used, where P is the crack initiation load and (a/b) is the specimen taper ratio.
Donaldson[26]	90° CN	G_{Ic} was obtained through calculation of $K_{Ic} = \sigma_{NC}\sqrt{\pi a}$, where a is the half crack length, and σ_{NC} is the far-field stress normal to the crack. G_I was associated with K_I via an equation containing the orthotropic material properties.
Whitney & Browning[22]	90° CN	G_{Ic} obtained through calculation of $K_{Ic} = \sigma_n^\infty\sqrt{\pi(a+a_0)}$, where a_0 is an inherent flaw determined from the average stress criterion, σ_n^∞ is the notch strength of plate of infinite extent. σ_n^∞ was obtained from σ_n, the strength of a tensile coupon with finite width, via an isotropic width correction relationship. G_{Ic} was obtained from K_{Ic}. G_{Ic} obtained was higher than values obtained through 0° DCB tests.

Wang *et al.*[31]	DEN	G_I available at notch-tip calculated by a two-dimensional plane-stress finite element routine.
Bradley & Cohen[16]	CT	Displacement–load relationship monitored. Tests conducted in displacement control to allow stable crack growth. Standard linear fracture mechanics relationship $G_I = (P^2/2B)(\partial C/\partial a)$ used.
O'Brien[23,24]	EDT	Laminated plate theory and simple rule of mixture for stiffness loss due to delaminations were used to derive $G_c = \frac{\epsilon_c^2 T}{2}(E_{LAM} - E)$, where E_{LAM} is the stiffness of plate calculated from laminated plate theory, E is the stiffness of partially delaminated laminate, ϵ_c is the critical strain at the onset of delamination, and T is the laminate thickness.
Whitney & Knight[25]	Modified EDT	Specimens had free-edge starter cracks of length a embedded in the centre plane of specimens. Classical laminated plate theory invoked to obtain $G_{Ic} = h\epsilon_c^2(E_x - E^*)$, where h is the half thickness of laminate, ϵ_c is the critical strain, and E_x and E^* are modified inplane stiffness terms. E here is not the same as O'Brien's E^*. Values of G_{Ic} obtained slightly lower than those from unmodified EDTs.
Whitney & Browning[22]	EDT	Laminated plate theory used to derive $G_{Ic} = h\epsilon_c^2(1 - E_1^*/E_1)\bar{E}_1$ where h is half laminated thickness, ϵ_c is the critical strain at delamination, E_1 is the laminate modulus in the load direction, E_1^* is an effective laminate modulus for delaminated composite, and $\bar{E}_1$ is the experimentally determined value of E_1. Both E_1 and E_1^* were calculated from laminated plate theory. If $\bar{E}_1 = E_1$, then the expression for G_{Ic} would be the same as O'Brien's expression.

TABLE 2

Comparison of Experimental Values of G_{IIc} from the Literature

Reference	*Description of test method and analysis*	*Material*	$G_{Ic}(J/m^2)$
Wilkins *et al.*[18]	CLS (Cracked Lap Shear) specimens: Relative proportions of G_I and G_{II} obtained through finite element analysis. Proportion of Mode II, G found to be approx. 75% of total energy release rate.	T330/5308	154
Ramkumar & Whitcombe[19]	CLS specimens. The embedded delamination did not always propagate along the midplane, but crossed over to the adjacent interface. A simple strength-of-materials analysis of CLS specimen to account for the change in stiffness due to the crossover was used to compute the change in compliance with crack length, dC/da. A geometrically non-linear finite element analysis was performed to determine the relative proportions of G_I and G_{II} at the critical load. G_c was obtained experimentally. Values of G_I, G_{II} and G_{Ic} were substituted into three failure criteria and G_{IIc} obtained: (a) $G_I/G_{Ic} + G_{II}/G_{IIc} = 1$, (b) $(G_I/G_{Ic})^2 + (G_{II}/G_{IIc})^2 = 1$, (c) $(G_I/G_{Ic})^2 + (G_{II}/G_{IIc})^2 + (G_I/G_{Ic})(G_{II}/G_{IIc}) = 1$	T300/5208	876, criterion (a) 456, criterion (b) 643, criterion (c)
Russell & Street,[20,27]	ENF (End Notched Flexural) specimens. Specimens were subjected to three point flexural loading. The Teflon starter notch tip was placed midway between two loading pins. Beam theory analysis was used to obtain dC/da to be substituted into the expression for G_{IIc}, yielding $G_{IIc} = 9a^2/P^2 C_b/2b(2L^3 + 3a^3)$, where C_b is the compliance due to bending.	AS1/3501–6 HMS/3501–5 AS4/2220–3	452 ± 6 152 ± 11 750 ± 25

Chatterjee *et al.*[32]	Three Point Bend Test specimens. Implanted defects in the form of two-ply Teflon delaminations were located in the region of compressive flexural stress. A computer code was used to solve the problem of disbonds between two piles. Some of the specimens were fatigued to produce sharp crackfronts.	AS1/3501–6	452 ± 214 Blunt crack tips 950 ± 175 Sharp crack tips
Donaldson[26]	Modified Three Rail Shear Test. This was not a delamination test, but an in-plane crack was cut into an unidirectional laminate parallel to the fibre direction. Earlier finite element analysis by Lakshminarayana showed a uniform state of in-plane shear stress throughout the central region of the specimen. This shear stress was computed via $\tau_c = P_c/2tl$, where P_c is the critical applied load, t and l are the specimen thickness and length respectively. The stress intensity factor $K_{IIc} = \tau_c\sqrt{\pi a}$, where a is the half crack length, was computed and the value of G_{IIc} thus calculated.	T300–1034C	506
Jurf & Pipes[29]	Modified Arcan Test on Single Edge Notch specimens. The fixtures enabled mixed-mode loading of the specimen by varying the angle of loading, α. The stresses applied to the specimen were simply $\sigma = (P \sin \alpha)/A$, for the normal stress, and $\tau = (P \cos \alpha)/A$ for the shear stress, were A was the area of specimen through which the load was transmitted. The specimens were 91 ply thick, with teflon inserts, and where adhesively bonded to the aluminium Arcan fixtures. For Mode II loading, $\alpha = 0°$. Some specimens failed at the metal–composite adhesive bond. K_{IIc} was obtained and related to G_{IIc}		

2.2 Mode II characterisation

The situation with Mode II delamination is unfortunately very different. There is wide variability of results for G_{IIc} in the literature for comparable composite systems. Table 2 shows that the variation of G_{IIc} obtained ranges from 154 to 1200 J/m^2. The experiments designed to measure G_{IIc} are often derived from tests to determine shear strength. Although the cracked lap shear specimen has been used to determine G_{IIc}, it is essentially a mixed-mode specimen. The proportion of G_{II} present in this test has been determined, through finite element analyses, to be between 75% and 80% of the total system energy release rate.[18,19] Since the cracked lap shear specimen is a mixed-mode specimen, the value of G_{IIc} is dependent on the failure criterion used[19] and the accuracy of the finite element analysis in determining the proportions of G_I and G_{II}.

Some researchers have attempted to obtain G_{IIc} more directly through pure Mode II tests. Russell & Street[20,27] have used an edge delaminated laminate specimen simply loaded as a beam specimen. Chatterjee's[32] three-point bend test was similar except that two delaminations were introduced instead of one and were located in the flexural compressive region of the specimen. Chatterjee, however, reported values of G_{IIc} about twice the value of those obtained by Russell & Street. Jurf & Pipes[29] modified the Arcan test fixtures and used them to test a single edge notch specimen. Because of the nature of the geometry required, this specimen had to be about 90 plies thick and bonded to an aluminium fixture. The main advantage of the Arcan test was that it enabled Mode I, Mode II and Mixed-Mode testing on a single specimen type. However, there was also a tendency for some specimens to fail at the fixture–specimen adhesive bond, thus rendering the test invalid. Donaldson[26] investigated Mode II failure of a notched in-plane crack through a modified three-rail shear test arrangement. It may be argued that since both the growth of in-plane delaminations and along-the-fibre cracks in unidirectional laminates are matrix-dominated, the three-rail shear test should yield a G_{IIc} that is dependent only upon the matrix material.

2.3 Discussion of test results

Much work remains to be done to properly characterise delaminations in composite laminates, particularly their Mode II and Mixed-Mode behaviour. Various failure criteria proposed have met with only limited success in explaining mixed-mode failure. Much evidence suggests that G_{IIc} is much higher than G_{Ic}. In mixed-mode fracture, the contributions of Mode I and Mode II components to the total energy release rate are complicated and not readily apparent. In particular, the individual quantities of G_I and G_{II} cannot necessarily be simply added together to obtain the total energy release rate. This is because, in three-dimensional delamination propagation, the growth of the crackfront is often non-self-similar, and furthermore the contribution of Mode II may become significant. The analyses and experiments carried out to date have concentrated mainly on the first two modes of failure, although for thicker non-unidirectional laminates, the contributions of Modes II and III become increasingly more significant than that of Mode I. Clearly a failure surface in three-dimensional (a failure envelope in two-dimensional) stress space needs to be developed. This goal is presently not yet attainable since there does not appear to be a set of consistent data for mixed-mode fracture from different investigations. In proposing failure criteria, expressions often used are of the form:

$$\left(\frac{G_I}{G_{Ic}}\right)^m + \left(\frac{G_{II}}{G_{IIc}}\right)^n = 1 \tag{1}$$

where m and n are constants. Although useful in curve-fitting the experimental data, such expressions lack feasible physical interpretations. Indeed the extension of such a law to three-dimensional failure, besides needing the value G_{IIIc}, has no physical backing. Any acceptable failure criteria based on energy release rates must have a sound physical basis and for the case of two-dimensional, self-similar growth, eqn (1) must reduce to the form $G_{tot} = G_I + G_{II}$. This philosophy has the following implications:

(1) A consistent test methodology must be developed for both G_{IIc} and G_{IIIc}, and must allow for the effect of local constraint.
(2) Less emphasis must be given to the curve fitting approach for mixed-mode fracture.
(3) A detailed three-dimensional analysis should be undertaken for each test method in order to determine the relative contributions due to G_I, G_{II} and G_{III}.

3 FATIGUE LIFE PREDICTION

3.1 Test results

Cyclic fatigue tests play a central role in the assessment of fatigue life and the specification of inspection intervals for metallic components. This has led to similar tests for composites with impact damage. It has been found that the

S–N curves generated from experimental data are often flat over a large range of cycles.[33–37] Tension–tension fatigue loading is considered to be less critical than compression–compression and tension–compression loading.[38–40] This paper will concentrate on the latter two loading cases. For compression–compression and tension–compression loading, the maximum residual compressive load as a fraction of the static failure load (S) decreases from 1·0 to typically 0·6 for N in the range 1 to 10^6, depending on the initial damage size.

The greatest rate of degradation appears to be during the early part of cyclic loading, for N up to about 100 cycles. As the number of cycles, and hence the damage size is increased, very little further residual strength degradation is observed. In fact, it has been observed that $S = 0{\cdot}6$ may be taken as the 'fatigue threshold' value, below which the component may be assumed to have 'infinite life'. Various researchers report threshold values of between 0·6 and 0·8.[11,33,35,41–44] *S–N* curves are invariably drawn from a best fit of limited experimental data often with large scatter and their shapes vary from straight slopes to some form of curve.

The differences in fatigue lives resulting from changes in maximum compressive stress or strain from different researchers are summarised in Table 3. From the figures shown in the table, it can be seen that an uncertainty in the magnitude of the applied maximum compressive stress, for example 10%, can lead to a hundred-fold difference in fatigue life. Of particular interest is the considerable scatter in the data recorded by Ryder *et al.*[35] and Potter[34] which show an order of magnitude of difference in the fatigue lives of specimens at a nominally constant stress level.

For service aircraft the in-flight loads on a component will not be accurately known. There is often a significant variation between aircraft, with the result that the stress levels will always be subjected to a degree of uncertainty. We have seen that a small change in stress can result in a very large change in fatigue life. This implies that laboratory testing cannot be expected to provide accurate estimates of the fatigue lives of impact damage laminates and therefore cannot be expected to provide accurate information on inspection intervals. If this is true then the rationale for conducting such fatigue tests becomes questionable. Indeed, there are two natural consequences of this phenomenon:

(1) During the designing of a component, the stresses should be kept below the threshold value, thus eliminating the possibility of a fatigue failure due to BVID.
(2) If an aircraft is already in service, and the design procedure had not included step (1), then there is little point to fatigue testing of impact damaged laminates. In this case, the emphasis for research and development should be moved towards:

TABLE 3
Summary of Experimental Data on *S–N* Curves Derived from the Literature

Reference	*Maximum compressive strain range*	*Maximum compressive stress range*	*Range of cycles to failure*	*Comments*
Rosenfeld & Gause[11]	0·0024–0·003 $\Delta\epsilon = 0{\cdot}0006$		10^3–10^6	Laminated plate specimens. Delamination area 12·9 cm^2, 16·4 J impact, $R = -\infty$
Demuts *et al.*[42]	Strain at 0·003		10^4–10^6	Multirib specimens, AS6/2220–3. 136 J impact, $R = 10$.
Demuts & Horton[33]	$S = 0{\cdot}6$–$0{\cdot}7$		5×10^4–10^6	3-stiffener panel, AS6/220–3. 136 J impact, $R = 10$.
Walter *et al.*[43]	$S = 0{\cdot}28$–$0{\cdot}30$		10^4–10^6	Laminated plate specimens, T300/5208. 6·2 J impact, $R = -1{\cdot}0$.
Stellbrink & Aoki[41] Stellbrink[44]		200–250 MPa $\Delta\sigma = 50$ MPa	0–10^5	Laminated plate specimens, T300/CODE69. Impacted with ball of mass 255 g and radius of 5 mm at velocity of 6·15 m/s. $R = -1{\cdot}0$.
Potter[34]		Stress level fixed at 206 MPa	622–8242	Tapered-thickness specimens, Fibredux XAS/914·5 J impact, $R = 10$.
Ryder *et al.*[35]		Stress level fixed at 241 MPa	2×10^4–10^6	24-ply damaged hole specimens. $R = -1{\cdot}0$.
		Stress level fixed at 152 MPa	2×10^4–5×10^5	32-ply damaged hole specimens. $R = -1{\cdot}0$.

(a) Static strength testing to determine critical damage size and loads, allowing for representative service conditions; and
(b) developing a rational reject or repair criterion for the component which includes a methodology for determining inspection intervals.

Indeed, as a result of this work it is clear that further research is required in order to understand the physical reasons why the pronounced fatigue threshold exists. Such an explanation should involve the ratio of the relative length scales of the damage size to the laminate thickness and the far-field strain energy density relative to the critical strain energy density of the undamaged laminate.

3.2 Prediction of damage growth

In order to characterise the delamination growth using fracture mechanics techniques, several recent investigations (see Chatterjee[47]; Mohlin *et al.*[48]; O'Brien[49]; Ramakumar & Whitcomb[50]; Wilkins *et al.*[54]) have adopted a growth law relationship of the type $(\mathrm{d}a/\mathrm{d}N) = c(\Delta G)^n$ or $(\mathrm{d}a/\mathrm{d}N) = c(G)^n$ where G is the mixed-mode delamination fracture energy, c and n are constants determined from experiments. Table 4 gives the value of the exponent n as predicted in several studies for different modes, that is Mode I, Mode II, and Mixed-Mode. At this point it is important to note that the values of the exponent n given in Table 4 are obtained by fitting a suitable curve to the experimental data. Hence, in order to get a good correlation for predicting the growth of a delamination, it is very important to obtain experimental data which are accurate and realistic. Table 4 reveals a large variation of n for Mode I loading and a relatively small variation for Mode II loading. Table 5

TABLE 4
Values of the Exponent 'n' in the Growth Law $(\mathrm{d}a/\mathrm{d}N) = c(G)^n$ as Predicted by Various Studies for Mode I, Mode II, and Mixed-Mode Cases

Reference	*Exponent 'n'*	
	Mode 1	*Mixed-Mode or Mode II*
Wilkins *et al.*[54]		
Wang *et al.*[51]	20–30	7–8 (Mixed-Mode)
Ramkumar & Whitcomb[50]	8·02–10·08	6·07 (Mixed-Mode)
O'Brien[49]	3·11–3·5	—
Chatterjee *et al.*[47]	—	2·373 (Mode II)
Whitcomb[53]	15–20	—

TABLE 5
Values of the Exponent 'n' in the Growth Law $(\mathrm{d}a/\mathrm{d}N) = c(\Delta G_{\mathrm{II}})^n$ as developed by Russell & Street[52] in the Order of Increasing Toughness

Exponent 'n'	*Material*
5·79—AS1/3501–6	First generation high temperature graphite–epoxy
5·71—AS4/2220–3	Second generation high temperature graphite–epoxy
4·52—C6000/F155	An intermediate temperature rubber toughened graphite–epoxy
3·88—AS4/APC2	High temperature graphite–thermoplastic composite

gives the values of n for Mode II which are taken from the experimental work by Russell & Street.[52]

Indeed when evaluating the delamination fracture energy in different modes (that is, G_{I}, G_{II} and G_{III}) by analytical and experimental means there will be inherent errors which are beyond control. As shown in section 2 the measured values for G_{Ic} vary from 80 to 240 J/m^2 depending on the matrix material and the nature of the fibre–matrix interface, and the variation of G_{Ic} in replicate measurements appear to be much greater than ±5%. For G_{II}, the situation is worse. The value of G_{IIc} depends very strongly on the test procedure and the data reduction scheme adopted (see section 2). Indeed Table 2 shows for T300/5208 a variation in G_{IIc} of between 154 and 876 J/m^2. With the difficulties involved in designing experimental methods for measuring pure Mode II fracture energy, it is often very difficult to measure G_{IIc} to an accuracy better than ±10%. We can expect similar errors to occur in measurements of G_{I} and G_{II} in fatigue tests. At this stage there has been no attempt to measure G_{IIIc} or to determine the error bounds on such a measurement.

The effect that these experimental uncertainties have on growth will now be examined. From Tables 4 and 5, take as a first approximation $n = 4$ for Mode II growth. Then the 10% uncertainty in G_{II} gives rise to 40% uncertainty in $\mathrm{d}a/\mathrm{d}N$. This situation is similar in Mode I. Although the error in G_{I} is smaller, say 5%, the exponent n is larger, typically 8 (see Table 4). This makes it pointless to use these growth laws in a predictive capacity.

An accurate prediction of delamination growth is very much dependent on the accuracy of measuring the values of delamination fracture energy during experiments. Furthermore, the scatter in the fatigue test data for composites gives rise to an additional uncertainty when it comes to applying this growth law in a service situation.

With the prevailing uncertainty in accurately measuring the value of Mode II delamination fracture energy, one questions whether it is worth putting any more effort into conducting extensive fatigue studies in order to predict delamination growth under cyclic loading.

The majority of the tests to date have concentrated on what was thought to be pure Mode I or Mode II growth. However, the growth of impact damage involves all three components. These conclusions are immediately apparent:

(a) A means of characterising Mode III growth is required.
(b) If the uncertainty in measuring G_{II} for a structural component cannot be overcome, then alternative ways of characterising the growth of delamination need to be considered.

3.3 Generalisation of the growth laws

There are two methods currently used to generalise the growth laws developed for pure Mode I or Mode II delamination growth. These approaches can be written as follows:

$$\frac{da}{dN} \propto G^n \tag{2}$$

$$\frac{da}{dN} \propto f_1(G_I) + f_2(G_{II}) \propto \lambda_1 G1^{n1} + \lambda_2 G_{II}^{n2} \tag{3}$$

where G is the total energy release rate with components G_I and G_{II} whilst n, $n1$, $n2$, $\lambda 1$, and $\lambda 2$ are experimentally determined constants. Equation (3) is empirical in nature and can be adjusted until it fits the data points for a series of Mode I and Mode II tests. Unfortunately it is entirely non-physical. It infers that the structure can subdivide the energy, i.e. G, available for crack growth into its various components G_I and G_{II} and then use these components in different ways. This is clearly not possible as it infers that the structure has an ability to think. There is therefore a significant risk in using this approach in structures where growth is neither pure Mode I nor pure Mode II. Whilst eqn (2) has a more physical interpretation, it must be remembered that G is really a vector quantity and does not equal $G_I + G_{II}$ unless growth is self-similar. We have also seen that the exponent n varies with the nature of the growth (see Table 3). For mixed-mode growth this infers that n will need to be separately determined. Indeed this cannot be totally achieved experimentally since for a general problem G cannot be measured from the movement of the load points. A combined experimental and finite element investigation is therefore needed for any particular problem.

3.4 Summary

From a brief review of studies conducted to predicted delamination growth in composites under cyclic loading it can be seen that it is very difficult to predict delamination growth. Furthermore, due to inaccuracies involved in measuring the values of Mode II and Modĕ III delamination fracture energy, which in turn govern the delamination propagation under cyclic loading, current growth laws may be of little use under service conditions.

To aid in the development of a method for predicting the delamination growth in composites, the experiments conducted for this purpose should be aimed at:

(a) Minimising the errors involved in measuring the value of delamination fracture energy.
(b) A reduction in the scatter.
(c) Development of alternative approaches.

4 RESIDUAL STRENGTH

4.1 Fracture mechanics concepts

There are two fracture parameters which are widely used to predict the residual strength of delaminated composite laminates. These are:

(i) The energy release rate approach.
(ii) The strain energy density approach.

4.1.1 The energy release rate approach

For Mode I self-similar crack growth of a through crack in the absence of body forces, the energy release rate G[55], can be written as:

$$G = \lim_{\epsilon \to 0} \int_{\Gamma_\epsilon} \left(W n_1 - t_i \frac{\partial u_i}{\partial x_1} \right) \mathrm{d}s \tag{4}$$

where W is the energy density, Γ_ϵ is a vanishing small closed path around the tip with normal n and t_i are the components of the traction vector on the path.

Here W is defined as:

$$W = 1/2 \epsilon_{kl} C_{ijkl} \epsilon_{ij} - \beta_{ij} \epsilon_{ij} (T - T_0) - \phi_{ij} \epsilon_{ij} (M - M_0) + C_1(T, M) \tag{5}$$

where C_{ijkl} is the stiffness tensor, β_{ij} and ϕ_{ij} are related to the coefficients of thermal and moisture expansion and C_1 is a function of temperature and moisture. Let us now consider the integral J_S which we will define as:

$$J_s = \int_{\Gamma_s} \left(W n_1 + t_i \frac{\partial u_i}{\partial x_1} \right) \mathrm{d}s \tag{6}$$

where Γ_s is the external boundary of the body. There is often a tendency to drop the subscript s and refer to J_s as J. Using Green's theorem it follows that:

$$G = J_s - \int_{V_S - V_\epsilon} \left[\frac{1}{2} \epsilon_{ij} \frac{\partial C_{ijkl}}{\partial x_1} \epsilon_{kl} - \epsilon_{ij} \frac{\partial}{\partial x_1} \left(\beta_{ij}(T - T_0) + \phi_{ij}(M - M_0) \right) + \frac{\partial C_1}{\partial x_1} \right] \mathrm{d}V \tag{7}$$

Thus J_s will not equal the energy release rate G unless the area integral vanishes. At constant moisture and temperature J_s may be equal to G. However, in general the area integral will be non-zero and J_s, which is measured experimentally from the movement of the load points,[56] will not equal G.

In service aircraft heating is often localised and the moisture content varies. This will produce a spatial variation in the tensor C_{ijkl} with the result that the area integral will in general be non-zero. Consequently when designing laboratory tests care should be taken to reproduce the near tip stress, strain, temperature and moisture fields rather than reproducing the 'global behaviour'. This is particularly true if the aim of the test is to establish such quantities as the critical damage size or the maximum permissible load.

For a three-dimensional fracture problem the integral on the right-hand side of eqn (4) is no longer equal to the energy release rate and is referred to as T^*.

4.1.2 Strain energy density approach

In the strain energy density approach, failure is assumed to occur when the available energy density W_{av} at a distance r_0 in front of the delamination in the direction of growth reaches a critical value W_c. The value W_c is dependent both on the values of $\mathrm{d}V(= \epsilon_{11} + \epsilon_{22} + \epsilon_{33})$ and $\mathrm{d}A$, the change of area per unit area.

For thermomechanical problems

$$W_{av} = 1/2\sigma_{ij}\epsilon_{ij} - 1/2\sigma_{ij}\alpha_{ij}(T - T_0) - 1/2\sigma_{ij}\psi_{ij}(M - M_0) - W_f \tag{8}$$

where $\beta_{ij} = \alpha_{ij} C_{ijkl}$, $\psi_{ij} = \phi_{ij} C_{ijkl}$ and W_f is the energy density in the fibre.

As shown in Refs 57 and 58 the strain energy density approach has certain advantages over the energy release rate approach. For self-similar growth both approaches give residual strengths in good agreement with experimental results.[57,58] However, for non-self-similar growth, energy release rate concepts must be used with caution.

It should be noted that finite element studies have shown that for delaminations in (a) composite laminates, (b) at step lap joints, or (c) at mechanically fastened composite joints, a stage is reached after which a significant increase in the size of the damage does not significantly reduce the residual strength. This significantly simplifies the methodology for estimating critical damage size.

5 BONDED REPAIR TECHNOLOGY

The repair concepts developed for composite structures make extensive use of classical lap joint theory. Unfortunately this theory is known to give erroneous results.[59,60]

When the structure being repaired is 'thick' classical theory dramatically over estimates the required overlap, or transfer, length. It also gives erroneous estimates of the load that can be successfully carried by the joint. In such circumstances a full three dimensional stress analysis is required along with a 'valid' failure criterion.

The energy density approach outlined in section 4.1.2 can also be used to design against delamination in adhesively bonded structures. Indeed recent work at the Aeronautical Research Laboratories, Australia,[61] has shown how this approach can be successfully used to design a one hundred and twenty ply boron epoxy laminate for application to the Wing Pivot Fitting of FIII-C aircraft in service with the Royal Australian Air Force.

6 CONCLUSIONS

This paper has attempted to outline the directions required for the through-life support of impact damaged composites. If a component is known to be fatigue prone, the emphasis in research should be on estimating a repair/reject criterion and on the development of advanced repair techniques. Research is also needed in the mechanics responsible for the observed fatigue threshold level.

At the moment repairs to thick monolithic structures with through-the thickness damage are carried out using mechanically fastened doublers. For impact damage this procedure may also be adopted. One major disadvantage with mechanically fastened repairs is that they provide a new path for moisture ingress. However, externally bonded scarf patches now offer a viable alternative. Perhaps the most important single requirement is to develop a need-to-repair criterion. The need for an efficient repair criterion is emphasised by results of research conducted under the Primary Adhesively Bonded Structure Technology (PABST) programme,[45] which showed that some bonded repairs made in the past were unnecessary. In some cases the repair served only to reduce the service life and had such flaws been unrepaired, the structures would have been just as strong and lasted much longer.

REFERENCES

1. Baker, A. A., Jones, R. & Callinan, R. J., Damage tolerance of graphite/epoxy composites. *Composite Structures*, **4** (1985) 15–44.
2. Badaliance, R. & Dill, H. D., Damage mechanism and life prediction of graphite–epoxy composites. *Damage in Composite Materials*, ASTM STP 775, 1982, pp. 229–42.
3. Schutz, D., Gerharz, J. J. & Alschweig, E., Fatigue properties of unnotched, notched and jointed specimens of a graphite–epoxy composite. Fatigue of Fibrous Composite Materials, ASTM STP 723, 1981, pp. 31–47.
4. Prakash, R., Significance of defects in the fatigue failure of carbon fibre reinforced plastics. *Fibre Science and Technology*, **14** (1981) 171–81.
5. Talreja, R., A conceptual framework for the interpretation of fatigue damage mechanisms in composite materials. *J. Composites Technology and Research*, **7** (1985) 25–9.
6. Harris, B., Fatigue and accumulation of damage in reinforced plastics. *Composites*, **8** (Oct.) (1977) 214–20.
7. Ramkumar, R. L., Compression fatigue behaviour of composites in the presence of delaminations. *Damage in Composite Materials*, ASTM STP 775, 1982, pp. 184–210.
8. O'Brien, T. K., Mixed-mode strain-energy-release rate effects on edge delamination of composites. *Effects of Defects in Composite Materials*, ASTM STP 836, 1984, pp. 125–42.
9. Jones, R., Broughton, W., Mousley, R. F. & Potter, R. T., Compression failures of damaged graphite epoxy laminates. *Composite Structures*, **3** (1985) 167–86.
10. Ramkumar, R. L., Effect of low-velocity impact damage on the fatigue behaviour of graphite–epoxy laminates. *Long-Term Behaviour of Composites*, ASTM STP 813, 1983, pp. 116–35.
11. Rosenfeld, M. S. & Gause, L. W., Compression fatigue behaviour of graphite–epoxy in the presence of stress raisers. *Fatigue of Fibrous Composite Materials*, ASTM STP 723, 1981, pp. 174–96.
12. Adsit, N. R. & Waszczak, J. P., Effect of near-visual damage on the properties of graphite/epoxy. *Composite Materials: Testing and Design, Fifth Conference*, ASTM STP 647, 1979, pp. 101–17.
13. Starnes, J. H., Jr, Rhodes, M. D. & Williams, J. G., Effect of impact damage

and holes on the compressive strength of a graphite/epoxy laminate, *Non-destructive Evaluation and Flaw Criticality for Composite Materials*, ASTM STP 696, 1979, pp. 145–71.

14. Williams, J. G., O'Brien, T. K. & Chapmen, A. J. III, Comparison of toughened composite laminates using NASA standard damage tolerance tests, *ACEE Conference on Composite Structures Technology, Seattle, WA*, 13–15 Aug., 1984.
15. Cardon, H. D., Impact dynamics research on composite transport structures, *ACEE Conference on Composite Structures Technology, Seattle, WA*, 13–15 Aug., 1984.
16. Bradley, W. L. & Cohen, R. N., Matrix deformation and fracture in graphite-reinforced epoxies. *Delamination and Debonding of Materials*, ASTM STP 876, 1985, pp. 389–410.
17. Chai, H., The characterization of Mode I delamination failure in non-woven, multidirectional laminates. *Composites*, **15**(4) (1984) 277–90.
18. Wilkins, D. J., Eisenmann, J. R., Camin, R. A., Margolis, W. S. & Benson, R. A., Characterizing delamination growth in graphite–epoxy. *Damage in Composite Materials*, ASTM STP 775, 1982, pp. 168–83.
19. Ramkumar, R. L. & Whitcomb, J. D., Characterization of mode I and mixed-mode delamination growth in T300/5208 graphite epoxy. *Delamination and Debonding of Materials*, ASTM STP 876, 1985, pp. 315–35.
20. Russell, A. J. & Street, K. N., Factors affecting the inter-laminar fracture energy of graphite/epoxy laminates. *Progress in Science and Engineering of Composites, ICCM-IV*, Tokyo, Japan Society of Composite Materials, 1982.
21. Aliyu, A. A. & Daniel, I. M., Effects of strain rate on delamination fracture toughness of graphite/epoxy. *Delamination and Debonding of Materials*, ASTM STP 876, 1985, pp. 336–48.
22. Whitney, J. M. & Browning, C. E., Materials characterization for matrix-dominated failure modes. *Effects of Defects in Composite Materials*, ASTM STP 836, 1984, pp. 104–24.
23. O'Brien, T. K., Analysis of local delaminations and their influence on composite laminate behaviour. *Delamination and Debonding of Materials*, ASTM STP 876, 1985, pp. 282–97.
24. O'Brien, T. K., Characterization of delamination onset and growth in a composite laminate. *Damage in Composite Materials*, ASTM STP 775, 1982, pp. 140–67.
25. Whitney, J. M. & Knight, M., A modified free-edge delamination specimen. *Delamination and Debonding of Materials*, ASTM STP 876, 1985, pp. 298–314.
26. Donaldson, S. L., Fracture toughness testing of graphite/epoxy and graphite/PEEK composites. *Composites*, **16**(2) (1985) 103–12.
27. Russell, A. J. & Street, K. N., The effect of matrix toughness on delamination: static and fatigue fracture under mode II shear loading of graphite fibre composites, Paper presented at NASA/ASTM Symposium on Toughened Composites, Houston, 13–15 March, 1985.
28. Byers, B. A., Behaviour of damaged graphite epoxy laminates under compression loading, NASA-CR-159293, Final Report, Dec. 1979, 1980.
29. Jurf, R. A. & Pipes, R. B., Interlaminar fracture of composite materials. *Composite Materials*, **16** (1982) 386.
30. Nicholls, D. J. & Gallagher, J. P., Determination of G_{Ic} in angle ply composites

using a cantilever beam test method. *J. Reinforced Plastics and Composites*, **2** (1983) 2.

31. Wang, A. S. D., Kishore, N. N. & Fong, W. W., On mixed-mode fracture in off-axis unidirectional graphite/epoxy composites. *Process in Science and Engineering of Composites, ICCM-IV*, Tokyo, Japan Society for Composite Materials, 1982.
32. Chatterjee, S. N., Pipes, R. B. & Blake, R. A. Jr, Criticality of disbonds in laminated composites. *Effects of Defects in Composites Materials*, ASTM STP 836, 1984, pp. 161–74.
33. Demuts, E. & Horton, R. E., Damage tolerant composite design development. In *Composite Structure—3*, ed. Marshall, I. H., Elsevier Applied Science Publishers Ltd, London, 1985.
34. Potter, R. T., The interaction of impact damage and tapered-thickness sections in CFRP. *Composite Structures*, **3** (1985) 319–39.
35. Ryder, R. J., Lauraitis, K. N. & Pettit, D. E., Advanced residual strength degradation rate modelling for advanced composite structures, AFWAL-TR-79-3095, Vol. II, Tasks II and III, July 1981.
36. Dorey, G., Structural life predictions for fibre reinforced composite materials. ONR Workshop on Damage Tolerance of Fibre Reinforced Composite Materials, Glasgow, UK, 12 Sept., 1985.
37. O'Brien, T. K., Interlaminar fracture of composites. *J. Aeronautical Society of India*, **37** (1985) 61–9.
38. Black, N. F. & Stinchcomb, W. W., Compression fatigue damage in thick, notched graphite–epoxy laminates. *Long-term Behaviour of Composites*, ASTM STP 813, 1983, pp. 95–115.
39. Rosenfeld, M. S. & Huang, S. L., Fatigue characteristics of graphite–epoxy laminates under compression loading. *J. Aircraft*, **15** (1978) 264–8.
40. Ratwani, M. M. & Kan, H. P., Compression fatigue analysis of fibre composites, *NADC-78047-69*, Sept., 1979.
41. Stellbrink, K. K. & Aoki, R. M., Effect of defect on the behaviour of composites. *Progress in Science and Engineering of Composites*, ICCM-IV, Tokyo (1982).
42. Demuts, E., Whitehead, R. S. & Deo, R. B., Assessment of damage tolerance in composites. *Composites Structures*, **4** (1985) 45–58.
43. Walter, R. W., Johnson, R. W., June, R. R. & McCarty, J. E., Designing for integrity in long-term composite aircraft structures. *Fatigue of Filamentary Composite Materials*, ASTM STP 636, 1977, pp. 228–47.
44. Stellbrink, K. K., On the behaviour of impact damage CFRP laminates. *Fibre Science and Technology*, **18** (1983) 81–94.
45. Hart-Smith, L. J., The design of repairable advanced composite structures, Douglas Paper 7550, 14–17 Oct., 1985.
46. Byers, B. A., Behaviour of damaged graphite epoxy laminates under compression loading, *NASA-CR-159293*, Final Report, Jan. 1978–Dec. 1979, Aug. 1980.
47. Chatterjee, S. N., Pipes, R. B. & Blake, R. A., Jr, *Criticality of Disbonds in Laminated Composites*, ASTM STP 836, 1984, pp. 161–74.
48. Mohlin, T., Blom, A. F., Carlsson, L. F. & Gustavsson, A. I., *Delamination Growth in a Notched Graphite/Epoxy Laminate under Compression Fatigue Loading*, ASTM STP 876, 1985, pp. 168–88.

49. O'Brien, T. K., *Characterization of Delamination Growth in a Composite Laminate*, ASTM STP 775, 1982, pp. 140–67.
50. Ramkumar, R. L. & Whitcomb, J. D., *Characterization of Mode I and Mixed-Mode Delamination Growth in T300/5208 Graphite/Epoxy*, ASTM STP 876, 1985, pp. 315–35.
51. Wang, A. S. D., Slomania, M. & Bucinell, R. B., *Delamination Crack Growth in Composite Laminates*, ASTM STP 876, 1985, pp. 167–235.
52. Russell, A. J. & Street, K. N., The effect of matrix toughness on delamination: static and fatigue fracture under NASA/ASTM. *Symposium on Toughened Composites, Houston*, 13–15 March, 1985.
53. Whitcomb, J. D., *Strain Energy Release Rate Analysis of Cyclic Delamination Growth in Compressively Loaded Laminates*, ASTM STP 836, 1984, pp. 175–93.
54. Wilkins, D. J., Eisenmann, J. R., Camin, R. A., Margolis, W. S. & Benson, R. A., *Characterization of Delamination Growth in Graphite–Epoxy*, ASTM STP 775, 1982, pp. 168–83.
55. Atluri, S. N., *Computational Methods in the Mechanics of Fracture*, North Holland, Amsterdam, 1986.
56. Atluri, S. N., Nakaguki, M., Nishioka, T. & Kuang, Z. B., Crack tip parameters and temperature rise in dynamic crack propagation. *Engineering Fracture Mechanics*, **23**(1) (1986) 167–82.
57. Jones, R., Tay, T. E. & Williams, J. F., Thermo-mechanical behaviour of composites. In: *Composite Material Response: Constitutive Relations and Damage Mechanisms* (Ed. G. C. Sih *et al.*), Elsevier Applied Science Publishers, 1988, Chapter 3.
58. Janardhana, M. N., Brown, K. C. & Jones, R., Designing for tolerance to impact damage at fastener holes in graphite/epoxy laminates under compression. *Theoretical and Applied Fracture Mechanics*, **5**(1) (1986) 51–6.
59. Chow, C. L. & Woo, C. W., On flaw size and distribution in lap joints, *Theoretical and Applied Fracture Mechanics*, **4**(1) (1985) 75–82.
60. Paul, J. & Jones, R., Analysis of the double overlap fatigue specimen, A.R.L. Structures Report 402, April 1984.
61. Jones, R., Molent, L., Baker, A. A. & Davis, M. J., Bonded repair of metallic components: Thick sections. In: *Proc. Int. Conf. on Analytical Testing Methodologies for Design with Advanced Materials, ATMAM '87*, Montreal, Canada, August 1987 (Ed. S. V. Hoa, G. C. Sih & J. T. Pindera), Martinus-Nijhoff Publishers, The Hague, in press.

Composite Structures **10** (1988) 75–81

Carbon Fibre Composite on the Viggen Aircraft

Roger Stenberg

Dept. TFSUM, Saab–Scania AB, Fach 58188 Linköping, Sweden

ABSTRACT

The Saab–Scania Aircraft Company has manufactured aircraft for the Swedish Air Force since 1940, starting with the B-18 Bomber. Other famous aircraft are the J29 'Flying Barrel', the J35 'Draken' (today in service in Sweden, Denmark, Finland and soon in Austria) and the 'Viggen' family. In the beginning of the 1990s the multi-role combat aircraft JAS39 'Griffon' will be delivered to the Swedish Air Force. Saab–Scania also have a leading market position in the commuter airplane segment of 30–40 passengers, Saab 340.

At a request of the Swedish Defence Material Administration, Saab–Scania 1975 started a programme introducing new materials for aircraft production. This resulted in redesigning of some metal parts by using carbon fibre composite.

The programme gave a soft start in using the new material but has given a broad experience in both manufacturing and service areas. This experience is being used when developing new military and civil products.

1 GENERAL

The Carbon Fibre Composite (CFC) programme for the Viggen aircraft (Fig. 1) was started up in the mid-seventies, with the main purpose of the programme being to gain experience, for design and production as well as in service, for the next generation of Swedish military aircraft.

The parts originally made in aluminium that were initially selected to be redesigned to composite were a flap at the canard wing (Fig. 2), a landing

Composite Structures 0263-8223/88/$03·50

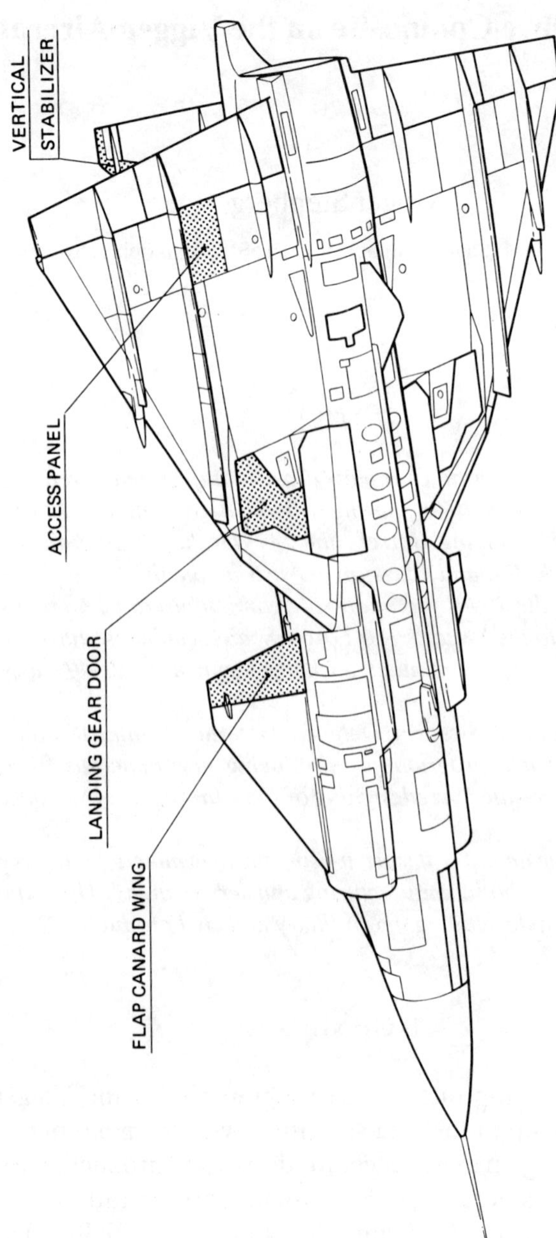

Fig. 1. Composite parts on the Viggen aircraft.

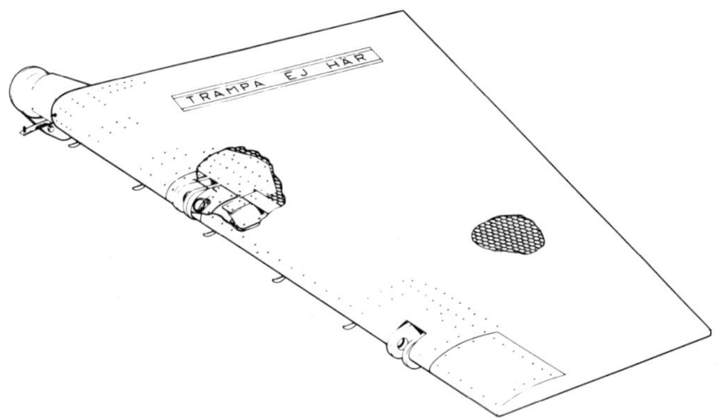

Fig. 2. Composite flap, canard wing.

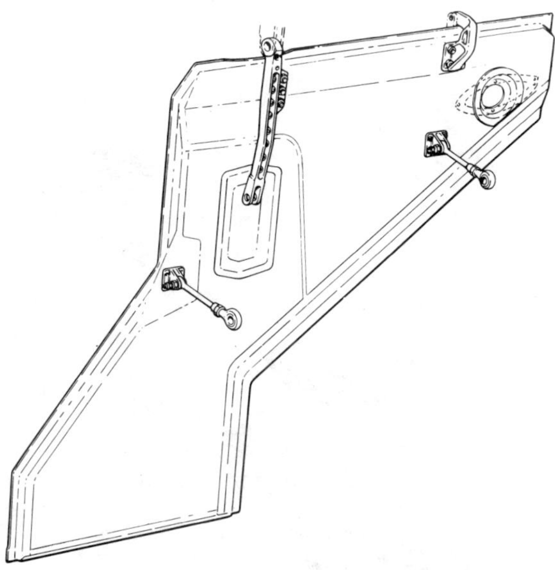

Fig. 3. Composite landing gear door.

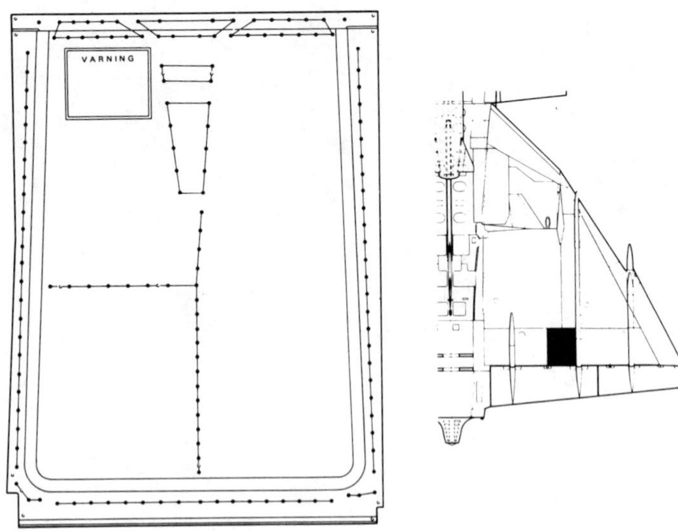

Fig. 4. Composite access panel.

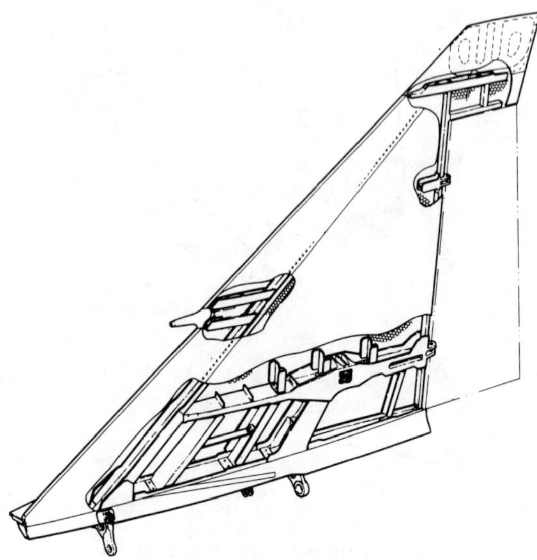

Fig. 5. Composite vertical stabilizer.

gear door (Fig. 3) and an access panel at the rear of the main wing (Fig. 4). For the last 20 reconnaissance aircraft the access panel became a standard unit, and the landing gear doors were standard units for the fighter version from the 16th aircraft onwards. The last selected part for the programme was the vertical stabilizer (Fig. 5) for the fighter version and the first units have recently been delivered to the Swedish Air Force. The programme is still ongoing and will hopefully offer results in the long run as good as they have been so far.

2 MAINTENANCE ACTIONS

Flap, canard wing: After every 50 hours flying a visual check of sealants and damage such as cracks and delaminations is made. After every 1300 hours flying an X-ray check is made to determine if moisture or water has caused any damage to the aluminium core.

Landing gear door, access panel and vertical stabilizer: After every 200 hours flying a visual check for damage such as cracks and delaminations is made.

In addition to these checks examination of the CFC parts has been made by Swedish composite specialists using more sophisticated equipment (e.g. ultrasonic).

3 IN-SERVICE EXPERIENCE

The CFC parts have now accumulated more than 30 000 flying hours with only minor damages apart from one exception. The exception being that the screws in the middle of the access panel tend to sink into the panel. The panel is a double skin panel with a core of aluminium. The aluminium core is enforced with a fill of epoxy resin with glass microballoons which was

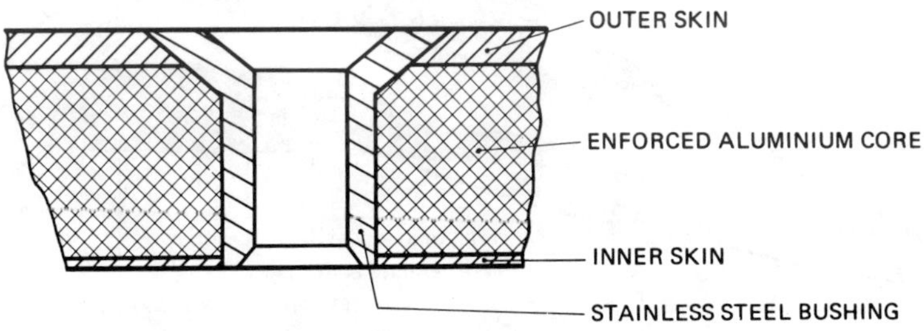

Fig. 6. Repair damaged screw hole with stainless steel bushing.

obviously too soft. The damage occurred when technicians used air-driven screwing machines. The panels were repaired by placing a stainless-steel bushing into the holes (Fig. 6).

The minor damages consist of smaller delaminations, three to four plies, around the screw holes and at the edges. Some damage has also occurred at the corners of the access panel. This damage is due to improper handling of the panel when removed from the aircraft. This minor damage is only cosmetically repaired to prevent growth and to stop moisture and water penetrating the laminate.

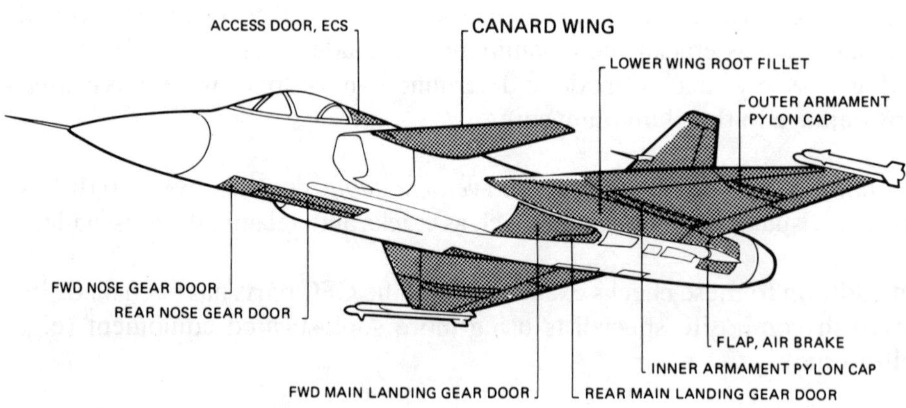

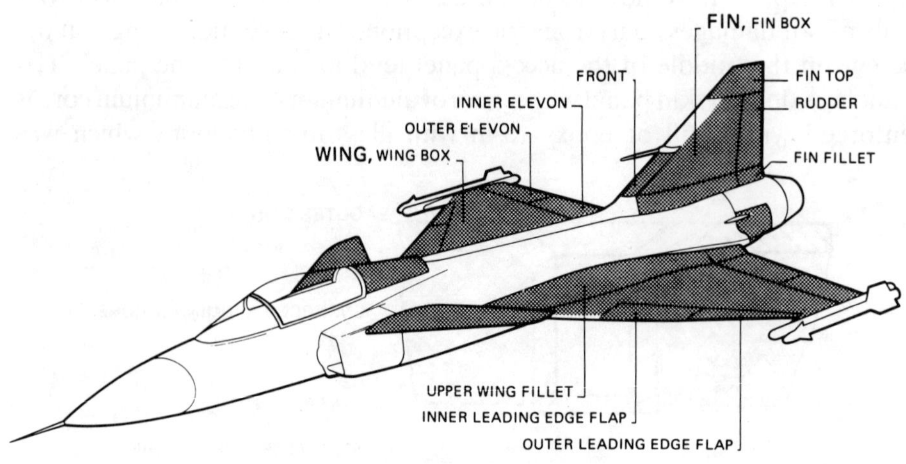

Fig. 7. JAS 39 Griffon composite parts.

4 AREAS OF CONSIDERATION

One of the most important areas requiring examination for the future is the development of a simple method of detecting delaminations in skin and substructure. Current methods of detecting delaminations are very expensive and time consuming, and require specialists to handle the equipment.

For Sweden, with its relatively small number of aircraft, it is essential that aircraft service and overhaul is made within as short a time as possible and by the technicians at the Air Force bases. For the next generation of military aircraft all with, for instance, composite wings it is most important that such methods should be developed.

Another area of concern is the development of proper repair methods and equipment for repairs that can be used and handled by technicians at Air Force bases. So far in Sweden all repairs are made by either the manufacturer or at depot level.

A group of specialists in charge of the Swedish Defence Material Administration with representatives from the industry and the Air Force are working on these problems in an effort to solve them before 1992 when the first JAS 39 Griffon (Fig. 7) will be delivered to the Swedish Air Force.

Composite Structures **10** (1988) 83–104

Damage Tolerance and Supportability Aspects of ARALL Laminate Aircraft Structures

J. W. Gunnink

Delft University of Technology, Faculty of Aerospace Engineering, PO Box 5058, 2600 GB Delft, The Netherlands

ABSTRACT

Comparison of ARALL laminates with other aircraft materials on a structural level shows that ARALL laminate is a very attractive material, especially for fatigue dominated structural parts. Investigation of this material in relation to the primary aircraft allowables gives some remarkable results. Fatigue and static testing of an ARALL F-27 lower wing panel confirms this.

INTRODUCTION

To prove the excellent behaviour of ARALL laminate material it is necessary to design, manufacture and test realistic components. First of all, aircraft design allowables are discussed in relation to the ARALL laminate material. This is followed by a general comparison of candidate aircraft materials with each other. Results of preliminary design studies on the outerskin of the lower wing of the Fokker F-27 are also shown. Finally the results of the ARALL F-27 lower wing panel are discussed, especially in relation to the design allowables and to the supportability of ARALL laminate structures.

PRIMARY DESIGN ALLOWABLES

It is common knowledge that for aircraft structural design several allowables are of primary importance: static strength (design limit and ultimate load), durability and damage-tolerance. These allowables can be found from the Federal Airworthiness Regulations (FAR) part 25 and also the Joint Airworthiness Requirements (JAR) 25, as requirements and design guide-

Composite Structures 0263-8223/88/$03·50

lines to achieve structural integrity. To obtain structural integrity it is necessary to determine as accurately as possible the load history of the structure as well as the design limit and the design ultimate loads. In this section the design allowables will be considered.

A new aspect, however, particularly relevant in the area of military aircraft and which can affect the design of aircraft structures significantly should also be considered. This aspect is called supportability. The possible impact of supportability on the design allowables will be discussed briefly. It is obvious that the abovementioned design allowables for ARALL laminates cannot be determined by the same factors used for both ordinary aluminium alloys and for composites.[1] The strength allowables for composites are primarily determined by the notch factor (holes, flaws, etc.) and environmental effects, whereas for ordinary aluminium alloys they are determined mainly by the material characteristics.

In many publications durability and damage tolerance are treated as separate items. Durability is in fact concerned with economic life and serviceability of the airframe, whereas the damage tolerance of the structure is defined as the ability of the structure to retain adequate strength and stiffness after some damage has occurred. Since 'damage' includes fatigue, corrosion as well as accidental damage, this means in general that the durability requirements also cover the damage tolerance assessments. In practice the durability 'allowable' is related to crack-initiation (fatigue) for aluminium alloys (in different environments), and to compression fatigue, holes and delaminations together with environmental effects for composites. On the other hand the damage-tolerance allowable for aluminium alloys is primarily dictated by crack-propagation, residual strength and inspectability. In the case of composite structures this allowable is dominated by residual strength, especially in compression loading of impacted structures. To define suitable allowables for ARALL laminates much research has been done and is still going on.[1–5] From the results so far it can be concluded that ARALL laminates behave in one respect as a metal and in another respect like a composite. The material also has its own characteristics, which cannot be related directly to either metals or composites. At the present time not all the data necessary to define the design allowables satisfactorily are available. However, with the available data, reasonable and acceptable allowables can be determined for the preliminary design of aircraft structures.

Static strength

The static ultimate allowable strength of ARALL laminates is, like composites, determined primarily by the notch factor, as shown in Fig. 1.

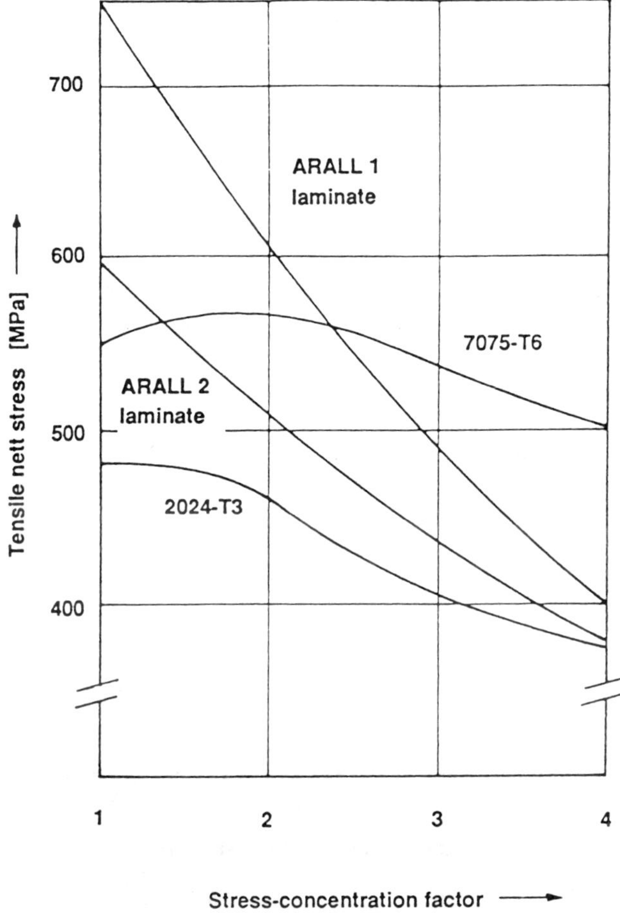

Fig. 1. Notch behaviour of materials.

With increasing notch factor the static strength of ARALL laminates decreases substantially; for instance for an open hole ARALL 1 laminates reduces to 65% and ARALL 2 laminates to 75% of the unnotched strength. The same behaviour, and the same order of magnitude, is observed from preliminary tests on low energy impacted ARALL laminates.[1] However, environmental effects, which are decisive for advanced composite materials, are virtually absent in tests on ARALL laminate specimens, even in very aggressive environments.[4] From the available data it is concluded that for the static tensile strength allowable the notch figures are the predominant factors for ARALL laminates. At present the static compressive strength allowables are not well enough defined. However, preliminary figures for the buckling behaviour and compressive yield strength indicate that the compressive buckling stresses are approximately 15% lower

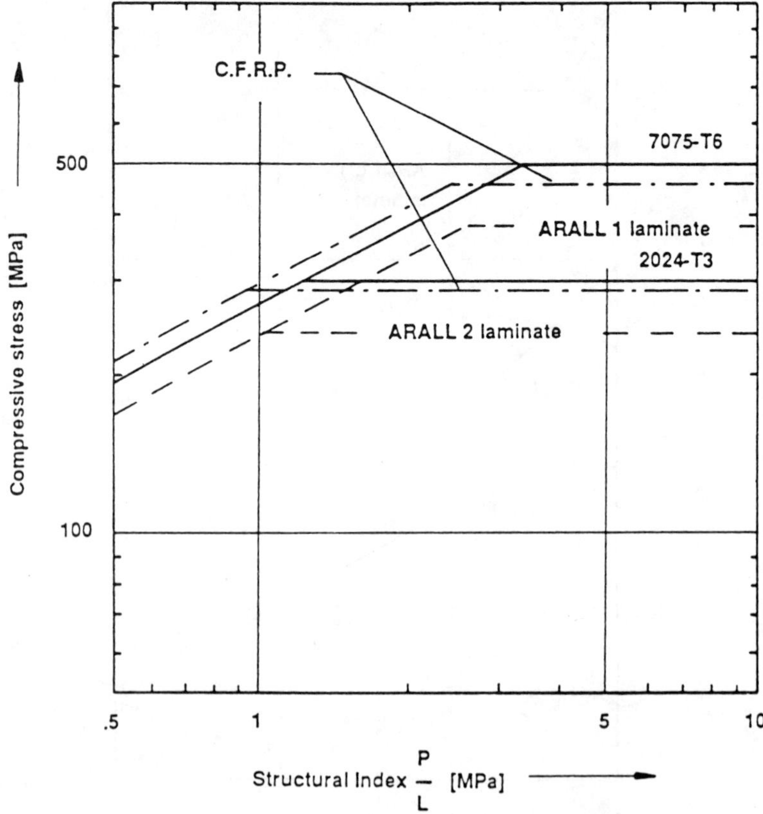

Fig. 2. Structural ultimate compressive strength.

compared to the related aluminium alloys, whereas the compressive strength is approximately 25% lower. In Fig. 2 the relationship between buckling, respectively compressive stress and structural index of stiffened plates, is given.

Durability

From extensive durability tests[4] it is found that different temperature ranges in combination with all kinds of possible aggressive environments (for the aircraft) show hardly any degradation in strength and stiffness (approx. 5%) even after exposure for more than 60 weeks.

For aluminium alloy, as mentioned, durability is directly related to its fatigue and crack-initiation behaviour. Fatigue cracking has been found to be the most prevalent form of degradation for aluminium structures. Environmental effects have a substantial and even dominant influence on it.

Crack initiation occurs in the outer aluminium layer of ARALL laminates at about the same number of load cycles compared to their related aluminium alloys. However, subsequent growth to macro-cracks does not occur in ARALL laminates, and neither does it occur at much higher fatigue stresses and much later than in the aluminium alloys. In most cases it is followed by complete crack arrest. Tests have shown that the maximum tensile stress of ARALL laminates is hardly degraded after fatigue loading (even in very aggressive environment) to more than three times an aircraft life. The reduction is of the order of 5–10%. It can be concluded from these tests that crack initiation is not a decisive factor influencing the durability of ARALL laminates. It will be shown in the next section that damage tolerance primarily dictates the durability of aircraft structures.

Damage tolerance

As mentioned above, damage tolerance is the ability of a structure to sustain loads after fatigue, corrosion and accidental damage until detected through inspection and repaired in such a way that the structure is again able to sustain the design ultimate load.

Damage tolerance consists of three aspects:[6,7]

—Residual strength: This defines the maximum damage, including the possibility of multiple cracks, that the structure can withstand under specified fail-safe load conditions.

—Crack propagation: This defines the time period in which a crack grows from a defined detectable length to the allowable length determined by the residual strength requirements.

—Damage detection: This defines inspection methods and intervals to ensure timely detection of cracks and other damage.

The relative importance of these aspects depends very much on the material used, the load level of the structure and the required life of the aircraft. Before outlining these aspects the different types of damage will be discussed. Very intensive fatigue and corrosion testing has been performed on ARALL laminates.[1,4] A fatigue crack starts from a stress concentration after the same number of fatigue cycles both in aluminium alloy and in ARALL laminates. However, the crack stays in ARALL laminates primarily in the outer layer, propagates very slowly and in most cases crack-stop is observed (Fig. 3). The crack length depends on the fatigue load level. For practical applications the crack length is found to be of the order of 2–5 mm. Preliminary residual strength tests of these fatigue panels (fibres still intact) show a very small reduction in strength compared with un-fatigued panels, of the order of 5–10% for both grades of ARALL laminate

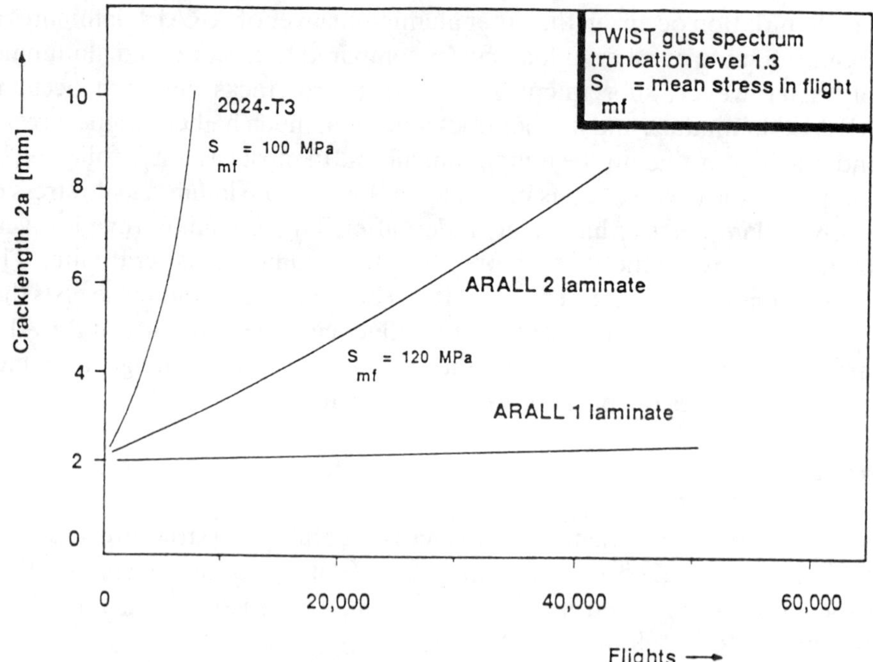

Fig. 3. Crack propagation rates in ARALL laminate materials.

materials (Fig. 4). Corrosion tests also show that ARALL laminates behave very satisfactorily and that the aramid layer acts as a barrier so that only the outer aluminium layer shows some pitting corrosion; ordinary aluminium alloys were fully corroded in the same test period. Reduction in static strength is minor, of the order of 5%. Fatigue testing of corroded ARALL laminates shows the same behaviour as for the uncorroded material.

The most critical type of damage seems to be artificial or accidental damage. Because of this type of damage the static strength of an unnotched specimen decreases substantially. Preliminary low impact energy tests on ARALL laminates, when the fibres are broken in the damaged area, shows a reduction in strength of 25–40% in comparison to unnotched specimens.[1] In fact this reduction is of the same order of magnitude or even slightly better than the already mentioned notch effect in ARALL laminates (Fig. 1). However, fatigue crack growth of these damaged ARALL laminate specimens shows the same behaviour as for the above described damage types.[1,8]

From the results so far it can be concluded that artificial or accidental damage is the most critical type of damage, from the residual strength point of view. In fact the crack growth rate for all types of damage shows about the

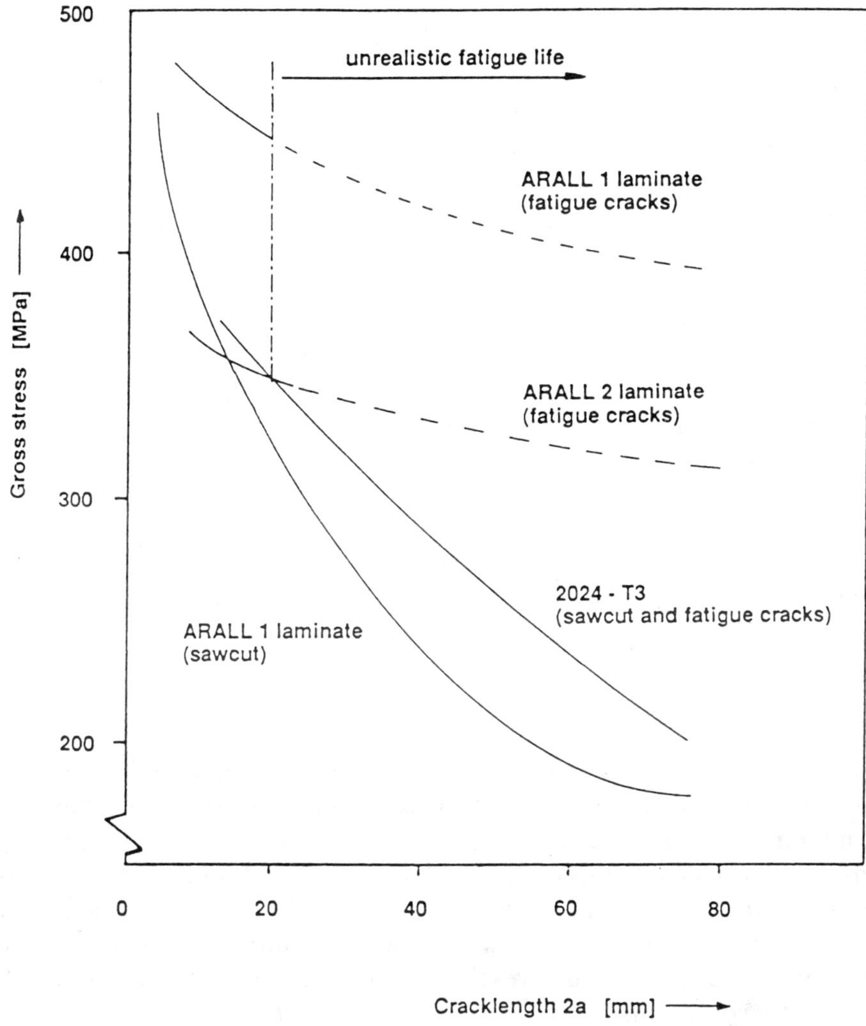

Fig. 4. Residual strength of unstiffened panels.

same kind of behaviour in practical types of application, crack extension of the order of 5 mm combined with a decreasing crack growth rate, and in most cases even crack arrest. The reduction in strength of fatigue specimens, compared to the initial (accidental, corroded, notched, etc.) specimens is very small (about 10%). Some preliminary conclusions can be drawn from this. First of all, crack propagation seems to be of little significance for ARALL laminate structures. Secondly, the residual strength of damaged ARALL laminate structures is important only in case of accidental damage. The static notch factor has such a significant influence on ARALL laminates that the small reduction due to fatigue crack growth by all kinds of damage seems adequately covered. In fact, the results so far indicate that the reduction in strength of structures with fatigue cracks is so small compared to unfatigued structures that (due to the static design allowable) the fatigued structures still can carry the ultimate design load. This means that ARALL laminate structures should be allowed to continue to fly with fatigue cracks (in which only the outer aluminium layers contain a crack) of the order of 5–10 mm.

On the other hand, due to the notch sensitivity of ARALL laminates, accidental damage is the decisive factor from a damage tolerance point of view, i.e. it gives in principle the maximum allowable initial crack length due to the impact of foreign objects. However, it must be realized that the corresponding load is a well defined residual strength load, mostly taken as 1·0 times limit load. This means that after detecting such a crack it must be repaired in such a way that the structure is capable of sustaining the design ultimate load.

Because accidental damage can only occur in specific areas, primarily the external parts of the structure, inspection of ARALL laminate structures seems to be much easier than for aluminium structures.

Supportability

Application of composite materials in primary airframe structures has revealed a lot of in-service problems. This has occurred especially in the area of military aircraft, which have the longest experience with primary composite structures. From these problems the question of supportability for composite structures arose. However, until now supportability has had no established definition. Probably the broadest description of supportability is 'The ability of airframe structures to meet the designed aircraft life within all operational conditions.'

In this definition supportability is closely related to the previously mentioned damage tolerance aspects. However, supportability also comprises the broad field of repairability. This often means in practice that supportability is directly related to maintainability and repairability. In fact, this shows the weakest link in the common usage of composite materials in primary aircraft structures. During the last two decades a tremendous effort has been put into optimizing composite aircraft structures to achieve structures with the least possible weight. Due to the nature of composite materials it is possible by this process to arrive at a tailor-made structure. However, investigation of repair techniques for these rather complex structures was mostly done in an *ad hoc* manner and seen as a subject of minor importance; maintenance people are also often not skilled enough to perform the tasks demanded.

If the area of supportability is only focused on maintainability and repairability the subject will become restricted; in fact dealing with composite structures like ordinary aluminium structures. Aluminium structures reduce, due to the nature of the aluminium alloys, the effects of mistakes made by the designer in a lot of cases. In this respect the low notch sensitivity of aluminium structures is very important. However, composite materials are notch sensitive. In general the sensitivity grows by increasing anisotropy of the laminates. For the tailor-made composite structures the notch sensitivity is in practice always higher than the least possible notch factor of composite laminates. This can result in a laminate lay-up which cannot be repaired in such a way that the structure is again able to sustain the design ultimate load. So the designer has to take account of this aspect, meaning that supportability has a direct impact on the design procedure. ARALL laminates are notch sensitive, but the sensitivity is somewhere between the notch sensitivity of aluminium alloys and composite materials. For ARALL laminate structures repairability can have an impact on the initial structural design.

OVERALL COMPARISON OF MATERIALS

In Table 1 the mechanical properties of some aircraft materials are given. For CFRP the unidirectional properties will reduce substantially due to the applied different fibre orientations in a structure. Even these reduced properties will not be the structural design allowables, which were discussed briefly in the previous sections. Environmental effects have a great influence on aluminium alloys as well as on CFRP. These effects together with fatigue cracking dominate aluminium structures. For CFRP, however, the low impact resistance, stress concentration sensitivity and bad repairability together with environmental effects are the essential considerations which will reduce the ultimate design strains significantly. According to today's practice the ratio of compressive strain of CFRP is in the order of 0·25–0·40%, and for the tensile strain in the order of 0·4–0·5%.[9] However, taking into account the total area of supportability, these figures can be lowered in some cases. Because fatigue cracking dominates aluminium aircraft structures their designed lifetime is an important factor when designing the structure. Simons *et al.*[10] have presented a general relationship between the ultimate gross area stress and the fatigue life for aluminium transport aircraft. It is obvious that by increasing lifetime the ultimate gross area stress reduces (Fig. 5).

As mentioned in previous sections the tensile design allowable for ARALL laminates is primarily the stress concentration sensitivity of the material. Due to the relative notch sensitivity of ARALL laminates,

TABLE 1
Mechanical Properties

Average properties		*ARALL 1 laminate*	*ARALL 2 laminate*	*7075-T6*	*2024-T3*	*CFRP UD*
Tensile ultimate	L[b]	772	703	572	455	1 240
Strength (MPa)	LT[c]	379	317	572	488	55
Tensile yield	L	655	351	510	344	NA
Strength (MPa)	LT	324	234	496	310	NA
Compressive yield	L	365	240	503	297	1 240[a]
Strength (MPa)	LT	379	270	524	338	207[a]
Elongation	L	0·6	1·3	11	18	NA
percentage	LT	6·3	12·4	11	18	NA
Elastic tensile	L	69·6	70·2	71	72·4	145
modules (GPa)	LT	52·1	52·9	71	72·4	12
Ultimate strain	L	1·7	2·5	12	19	0·9
percentage	LT	7·8	13·6	12	19	0·47
Blanking shear (MPa)		262	248	345	290	—
Density (g/cm^3)		2·29	2·29	2·78	2·78	1·6

[a] Compressive ultimate strength.
[b] L = in fibre (ARALL laminates and CFRP) or rolling (aluminium alloys) direction.
[c] LT = transfers to fibre or rolling direction.

compared to aluminium alloys, the level of experience of the designer plays an increasing role in aircraft structural design. It should also be noted, however, that the notch sensitivity of CFRP is far more pronounced. Together with test results of ARALL laminate structures this has resulted in the preliminary conclusion that a practical range of notch factor is 2·5–3·5. However, recent results of tests on large structural components have indicated that the upper part of this range is more suitable. Also, preliminary results of (impacted) ARALL laminate structures show that the repaired structures are covered by the upper range of notch factors, meaning that repairability will have a relatively small impact on the structural design.

Figure 5 plots the corresponding range of tensile strength together with the previously mentioned range for CFRP. In this figure the general comparison of aircraft materials, for lower wing application, shows the beneficial tensile strength of both grades of ARALL laminates over aluminium alloys as well as CFRP. Comparison of the compressive strength and loading shows a different behaviour. Taking into account what has been

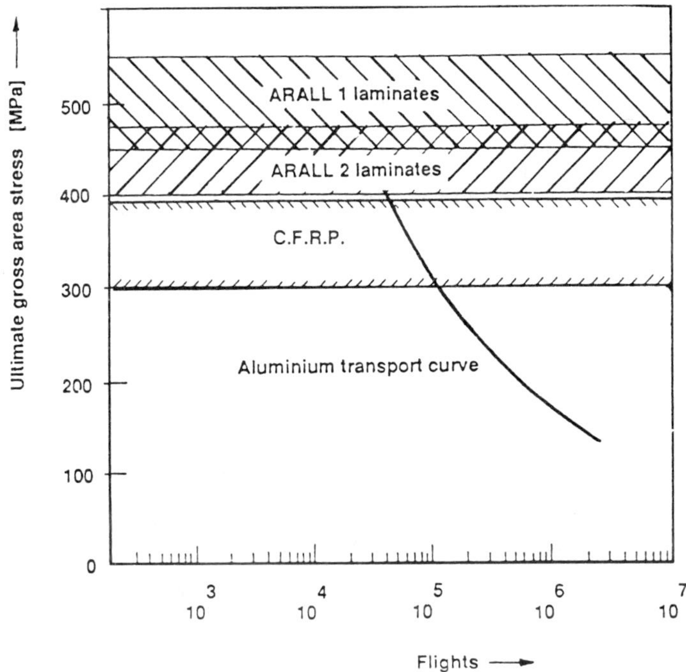

Fig. 5. Relationship between gross area stress and aircraft life.

previously stated, the maximum compressive stresses of both grades of aluminium alloys (7075-T6 and 2024-T3) are higher than the corresponding grades of ARALL laminates. The maximum compressive strength of CFRP depends very much on the allowable compressive strain. Due to its impact sensitivity the allowable compressive strain of CFRP in turn depends very much on the applied thicknesses. As Fig. 2 shows, the maximum compressive strength range of CFRP lies in the range of aluminium alloys and ARALL laminates. However, if the materials are compared to each other in relation to their density the results are more pronounced. Because aluminium alloys are (very) fatigue sensitive, and therefore (fatigue) life dependent, the comparison of fatigue and tensile dominated structural components has to be made for the designed lifetime of aircraft. A range of 30 000 to 100 000 flights is taken for this purpose. The fatigue–tensile structural efficiency curves (Fig. 6) show clearly that ARALL laminates are promising aircraft material for this type of application. The comparison also shows that CFRP and ARALL laminates are, from a structural viewpoint, genuine competitors. This means that other conditions, such as economics, also become important. However, this subject area is beyond the scope of this paper. A comparison of compression structural efficiency (Fig. 7) shows

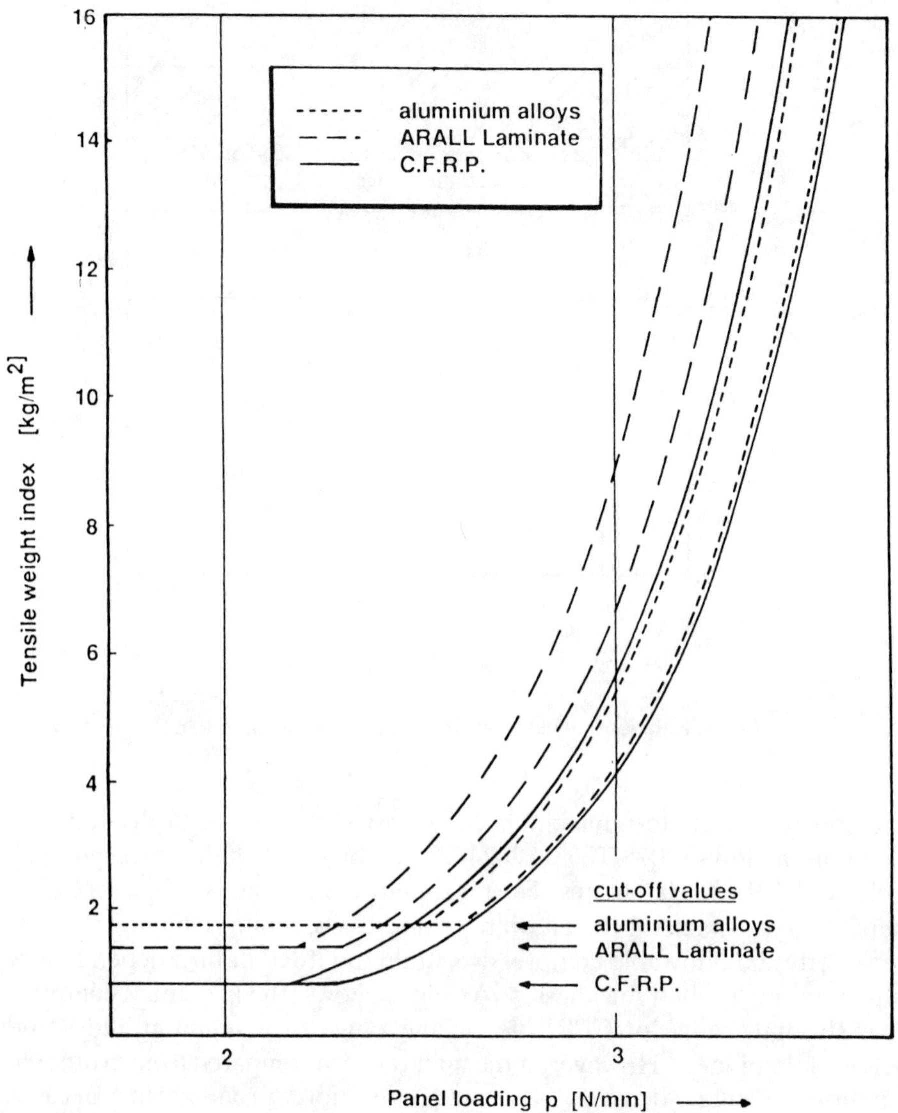

Fig. 6. Fatigue–tensile structural efficiency curves.

clearly that both grades of ARALL laminates and their corresponding aluminium alloys have about the same efficiency, whereas CFRP behaves in almost all aspects much better.

LOWER WING STRUCTURE OF THE FOKKER F-27

Alongside the development of the ARALL laminates, structural design studies were performed at the Delft University of Technology on the lower

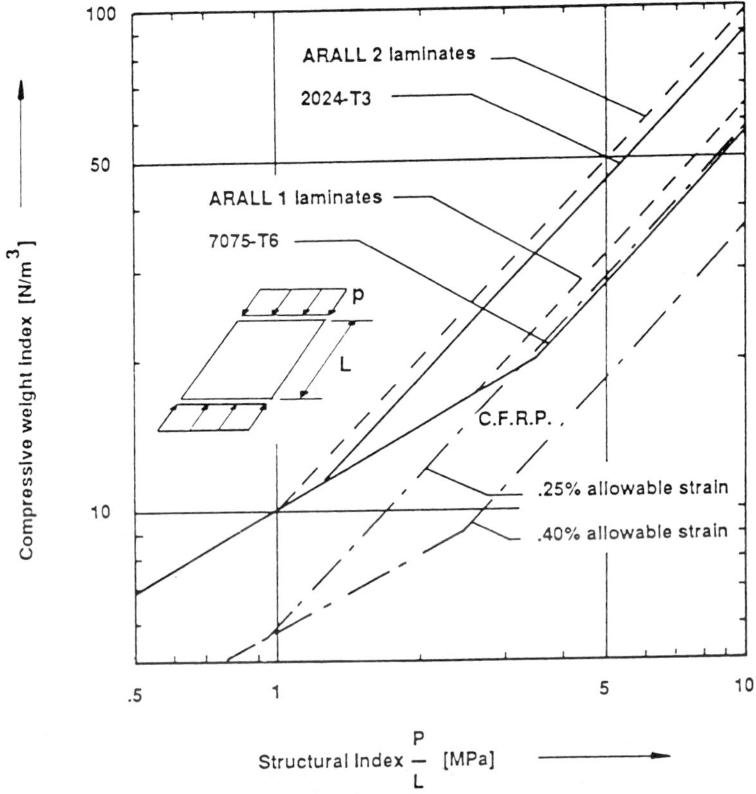

Fig. 7. Compression structural efficiency.

wing of the Fokker F-27 Friendship. This aircraft was chosen primarily because the structural lay-out as well as wing loading were readily available.[11] The lower skin of the outerwing of this aircraft was selected for design in ARALL laminates (Fig. 8). If the results are satisfactory this structural part can be replaced, due to its boundary conditions, quite easily by the designed ARALL laminate structure. However, it was the conviction of Delft University that good results from the designed ARALL structure must be proven through manufacturing and realistic testing of the structure. In this respect the selected structural part has its advantages. Since Fokker have tested extensively a significant part of the structure in fatigue loading as well as statically[12] it is only necessary to design in detail, manufacture and test the related ARALL laminate structure. As already mentioned, several design studies were performed during the development stage of the ARALL laminate.[13,14] These studies came to the same overall conclusion; ARALL laminates show significant weight improvement in the Fokker F-27 lower wing structure. A weight saving of approx. 30% is possible. The designs also provide another effect. The lower wing of small to medium size aluminium

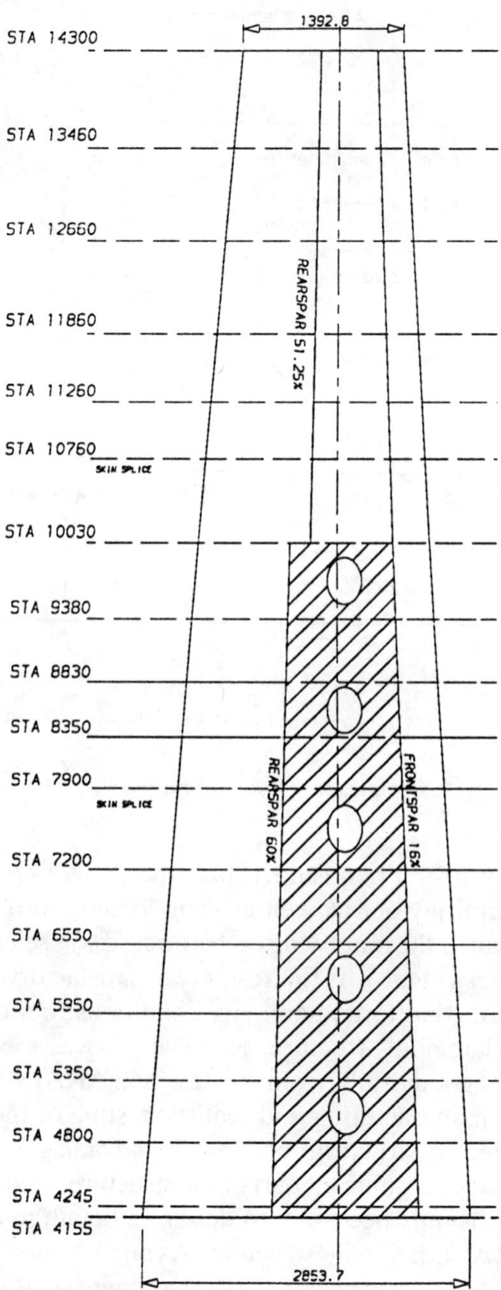

Fig. 8. Lay-out of the lower skin of the outerwing of the Fokker F-27.

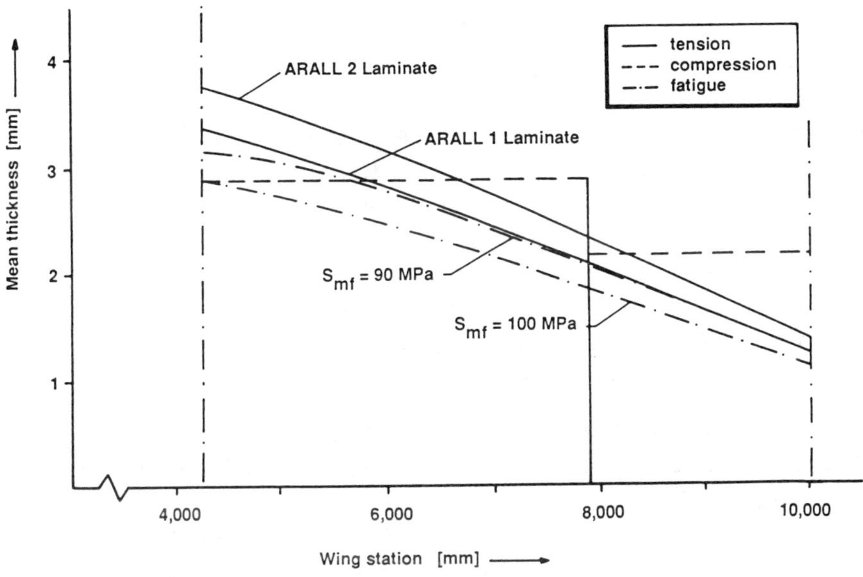

Fig. 9. Necessary thickness of the lower wing of the Fokker F-27.

transport aircraft are in general fatigue and tensile strength designed, i.e. the 1-g load level and the maximum positive gustloading are the decisive (load) factors. However, the ARALL laminate lower wing structure in the outboard direction is tensile critical but from station 6000 (ARALL 1 laminate) and station 7000 (ARALL 2 laminate) onwards, compression critical (Fig. 9). This means that the negative loading is the decisive loadcase for a major part of the ARALL laminate lower wing structure.

ARALL F-27 LOWER WING PANEL

The results of the preliminary structural design studies justified going ahead with a full-scale detailed design, and to carry out the testing of an ARALL lower wing panel. However, Delft University was not able to manufacture and fatigue test a panel of the required size (1·8 × 1·2 m). But in 1984 Fokker, convinced of the capabilities of the material, decided to participate in the ARALL lower wing panel project at Delft University, agreeing to manufacture and test the panel.[15] Delft University, as well as having overall responsibility for the project and some detailed testing, also undertook the design and analysis of the panel and the responsibility for the production drawings.

Originally in the test on the ARALL laminate structure, the following targets were set:

1. A 1-g stress level of 100 MPa.
2. The same fatigue loading as for the aluminium Fokker panels.
3. An average weight reduction of the ARALL laminate panels of about 25%, compared to the aluminium Fokker panels.
4. A crack free life of 45 000 flights.
5. A life without repairs of 75 000 flights.
6. An economic repair life of 90 000 flights.
7. A residual tensile strength of 1·1 times limit load.
8. A scatterfactor of 3 must be applied.

Later on some changes were made as a result of further evaluation of the material. For instance targets 1 and 2 can lead to a conflicting situation, as in fact occurs. However, to obtain the most realistic comparison, it was decided to give a higher priority to target 2. This resulted in the decision to fatigue test the panel up to the economic repair life including the scatter-factor of 3, or the ARALL F-27 lower wing panel has to be fatigue tested after 270 000 flights. The panel represents the wing location station 4155–5075 (Figs 8 and 10), which is the highest fatigue loaded area of the wing on the Fokker F-27. The outer wing/inner wing connection at station 4155 is a shear type connection. In designing the endfitting of the panel the same type of connection has been used. This major part of the ARALL F-27 lower wing panel has been extensively tested for fatigue.[16] A total of four ARALL laminate endfitting panels (Fig. 11) are fatigue tested. Two of them are finally tested until failure to determine the residual strength of these panels (Table 2). It was found after some small modifications that the behaviour of these panels was excellent. In fatigue after the equivalent of 300 000 flights only very small cracks in the outer aluminium layer of the stringers (at the end of the finger tips) were observed. Residual strength tests showed that the modified unrepaired panels did not fail at these cracks, but at the panel-rib connection. At this connection the panel has the smallest cross-sectional area and, due to the riveted connection of the rib-simulation to the panel, the panel has a notch factor which resulted in a reduction of the static strength of the panel below the residual strength of the fatigue cracked area. However, the residual strength was still about 5% higher than the required ultimate design strength. So the ARALL laminate endfitting panels behave satisfactorily.

These results have been incorporated in the design of the ARALL F-27 lower wing panel. Fokker has manufactured this panel according to the CAD-drawings at Delft University. It turns out that the panel has 33% less weight compared to the aluminium Fokker panels.

Fig. 10. ARALL F-27 lower wing panel.

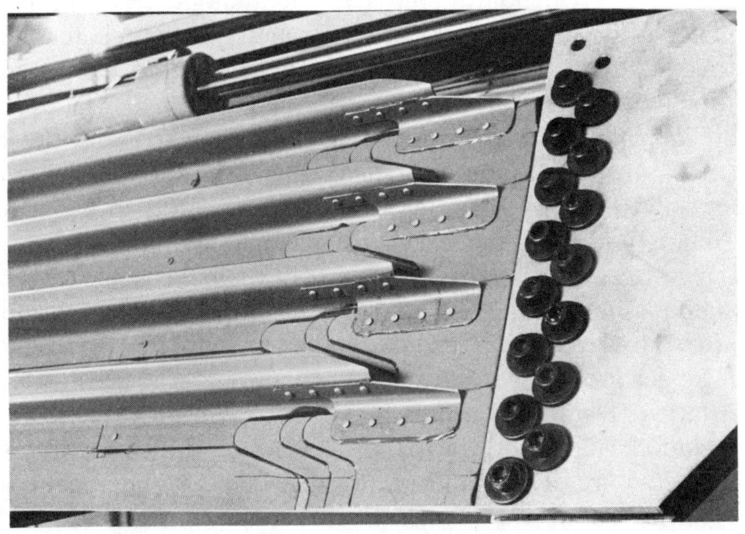

Fig. 11. ARALL laminate endfitting panels.

TABLE 2
Test Results of the ARALL Laminate Endfitting Panels

Endfitting panel	*Modification*	*Fatigue test*	*Residual strength test*
Panel 2 (ARALL 2 laminate)	—	300 000 flights 1-g stress level = 100 MPa	—
Panel 3 (ARALL 2 laminate)	Thickness reduction of the first doubler	270 000 flights 1-g stress level = 100 MPa	—
Panel 4 (ARALL 2 laminate)	Thickness reduction Change of geometry	270 000 flights 1-g stress level = 100 MPa	Yes
Panel 5 (ARALL 2 laminate)	As panel 4	90 000 flights 1-g stress level = 86 MPa 100 000 flights 1-g stress level = 100 MPa	Yes
Panel 6 (Aluminium)	As panel 2	86 000 flights 1-g stress level = 70 MPa complete failure	—

Testing of the ARALL F-27 lower wing panel showed some remarkable results. Due to severe bending in the cover area, fatigue cracks initiate after 20 000 flights at the longitudinal axis of symmetry in the rebate. This area was examined in more detail by finite element calculation and through this examination modifications have been performed (Fig. 12).[15] Static testing of the modified panel showed a reduction in bending stress at the critical location of about 50%. Continuing fatigue testing showed that the problems have been solved. Surprisingly the cracks did not grow in the remaining 250 000 flights, showing the excellent damage tolerance behaviour of ARALL laminates. During the fatigue test of the panel some small cracks were initiated, especially at the boltholes of the drainhole housing and in the rebate area. However, after the cracks had begun, they extended with decreasing crack growth rate and arrested after roughly 20 000 flights (Table 3). After fatigue testing the panel up to the economic repair life of 90 000 flights including the scatterfactor of 3, a residual strength test was performed. Fracture of the panel occurred at 1510 kN. The failure load of the panel turned out to be 1·42 times limit load.

It was remarkable that the lower wing panel did not fail at the fatigued areas (Fig. 13); in this respect the ARALL F-27 lower wing panel behaves in the same way as the ARALL laminate endfitting panels. However, the residual strength of the panel was 5% smaller than the design ultimate load.

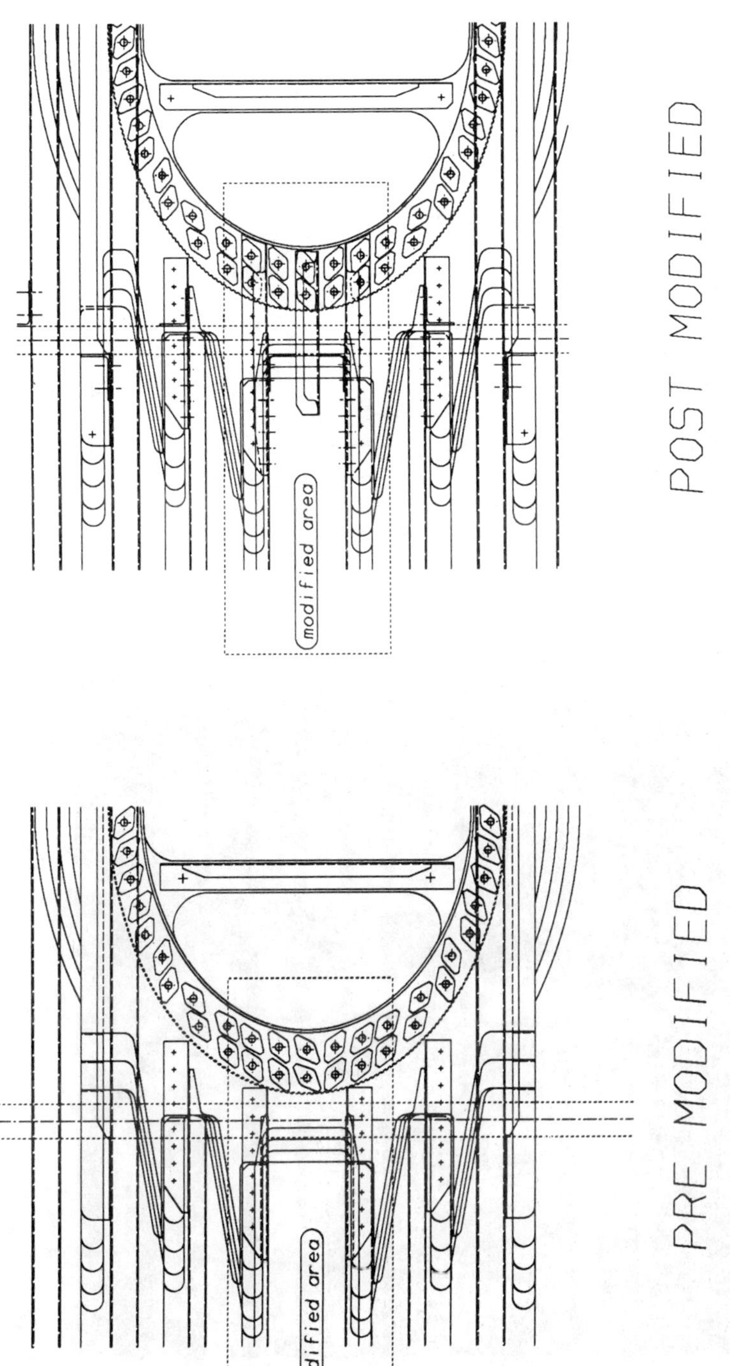

Fig. 12. Fatigue critical area.

TABLE 3
Overall Fatigue Results of the ARALL F-27 Lower Wing Panel

	Type of damage	*F-27 Panel (flights)*	*ARALL laminate panel (flights)*
A	Crack initiations in rebate (at transverse axis of symmetry)	10 000 (holes not cold expanded) 50 000 (holes cold expanded)	70 000
B	Crack initiation in rebate (at longitudinal axis of symmetry)	40 000	20 000[a]
C	Bolt holes at drain holes	30 000	80 000
D	Rivets and stiffeners at rib joints	45 000	>270 000
E	Fixation rivets for bonding	Does not exist	270 000
F	End of bonded doublers	30 000	270 000

[a] After design modification the crack growth stopped.

Fig. 13. Fracture of the ARALL F-27 lower wing panel in the residual strength test at 1·42 times limit load.

CONCLUSIONS

With the design and testing of the ARALL F-27 lower wing panel eight primary targets were set. The test results have shown that almost all targets were met. Only one target, concerning the crackfree life of 45 000 flights, needs some consideration. As previously mentioned, after 20 000 flights, some cracks occurred at the longitudinal axis of symmetry in the rebate of the manhole. This was due to the severe local bending in this area. However, after some re-analysis and redesign the problem was solved. Without repairing the cracks, the structure was able to continue fatigue testing up to 270 000 flights. During this extended fatigue testing hardly any crack growth was observed in the modified area. This shows the excellent damage tolerance behaviour of ARALL laminates. The test results of the panel support the conclusion that the notch sensitivity of ARALL laminates has a major impact on the design allowables of primary ARALL laminate structures. It shows by the design approach used that small fatigue cracks in the structure can be tolerated without repair, and in this condition the structure can still sustain the design ultimate load. Furthermore, the results indicate that the overall notch figure of 3 is difficult to meet in practical design. A more appropriate notch figure should be 3·5.

Some preliminary conclusions can be drawn also for supportability of ARALL laminate structures. In a lot of cases ARALL laminate structures behave more like aluminium structures than composite structures, which from a supportability viewpoint is favourable. There is one big positive exception, ARALL laminates are hardly fatigue sensitive. Another advantage of ARALL laminates over the aluminium alloys are their excellent durability, with hardly any degradation in strength or stiffness, even after a very long exposure in very aggressive environments. All these effects show the ability of ARALL laminates to be a viable supportable aircraft structural material. The recommendation that the structure should be designed with a notch factor of approximately 3·5 strengthens this statement. However, on-going research on repairability of ARALL laminate structures will have to prove this point.

REFERENCES

1. Vogelesang, L. B. & Gunnink, J. W., ARALL: a materials challenge for the next generation of aircraft. *Materials and Design*, **7**(2) (1986) Nov./Dec.
2. Gunnink, J. W., Verbruggen, M. L. C. E. & Vogelesang, L. B., ARALL, a light weight structural material for impact and fatigue sensitive structures. *Vertica*, **10**(2) (1986) 241–54.
3. Bucci, R. J., Mueller, L. N., Schultz, R. W. & Prohaska, J. L., ARALL

laminates–results from a cooperative test program, Alcoa Laboratories, presented at *Advanced Materials Technology 87, 32nd International SAMPE Symposium and Exhibition*, Anaheim, CA, April 1987.

4. Verbruggen, M. L. C. E., Aramid Reinforced Aluminium Laminates: ARALL adhesion problems and environmental effects, Report LR 504 and LR 505, PhD Thesis, Faculty of Aerospace Eng., Delft University of Technology, The Netherlands, 1987.
5. Anon., Alcoa ARALL laminate sheet, Alcoa Aerospace technical fact sheet, 1986.
6. Hall, J. & Goranson, U. G., Structural damage tolerance of commercial jet transports. *Boeing Airliner*, Jan.–Mar. (1984).
7. Schoevers, H., Development of structural maintenance, presented at the *IFA conference*, Amsterdam, Nov. 5, 1985.
8. Johnson, W. S., Impact and residual fatigue behaviour of ARALL and AS6/5245. *Composite Materials*, NASA TM 89013, Oct. 1986.
9. Riedinga, L. A. & Waraniak, J. M., Application and integration of design allowables to meet structural life requirements. ICAS-84-3.8.1, 1984.
10. Simons, H., Rhodes, J. E. & James, A. M., Strain barrier ultimate cut-off for aircraft structures, presented at *ICCM-3*, Paris, 25–30 August 1980.
11. Anon., Type-Record Fokker F-27 Friendship, Fokker Aircraft, Schiphol, The Netherlands, 1950.
12. v. d. Schee, P. A., Outerwing lowerskin fatigue testing as a part of the continuing airworthiness programme of the Fokker F-27, paper presented at the *10th ICAF Symposium*, Brussels 1979.
13. Heydra, J. J. & Venselaar, C. F., An ARALL lowerwing skin for the F-27 (in Dutch). Faculty of Aerospace Eng., Delft University of Technology, The Netherlands, April 1981.
14. Hegemans, G. & Racke, R., Application of ARALL sheetmaterial in the lowerwing of the Fokker F-27 (in Dutch). Faculty of Aerospace Eng., Delft University of Technology, Sept. 1982.
15. Gunnink, J. W. & v. d. Schee, P. A., Design of the ARALL F-27 lower wing fatigue panel, presented at *4th International Conference on Composite Structures (ICCS-4)*, Paisley, Scotland, July 1987.
16. van Veggel, L. H., Jongebreur, A. A. & Gunnink, J. W., Damage tolerance aspects of an experimental ARALL F-27 lower wing skin panel, presented at *14th ICAF Symposium*, Ottawa, June 1987.

Composite Structures **10** (1988) 105–108

Supportability of Composite Airframes: The Lavi Fighter Aircraft

A. Segal

Israel Aircraft Industries Ltd, Engineering Division, Ben Gurion International Airport, 70100, Israel

ABSTRACT

Israel Aircraft Industries (IAI) is the designer and manufacturer of the Astra executive aircraft, the Lavi fighter aircraft, and the pioneer mini RPV. Each of these products contains composite components. The experience gained to date from the use of composite materials has been quite good. Despite some problems during the development stages. Some of the problems encountered during the development of the Lavi are discussed.

INTRODUCTION

Supportability of a structure can be defined as the ability to function without excessive grounding for inspection and maintenance and with minimal financial outlay. The main factors affecting supportability of composite structures are discussed below, followed by a description of the IAI approach to supportability and the experience gained.

FACTORS AFFECTING SUPPORTABILITY

Materials and processes

Laminated composites in use today are very strong in the inplane direction, but very weak in the out-of-plane direction. One of the main weaknesses of the present generation of composite material systems is their limited toughness. Limited toughness is the main factor affecting the extent of damage caused by impact and when combined with the low out-of-plane

Composite Structures 0263-8223/88/$03·50

strength of the material, the result is a sensitivity to edge effects. Final material properties are sensitive to variations in production processes and are also affected by temperature and humidity.

Nature of typical damage

There are several types of defects and damage which may exist in composite structures. The two most important ones are delaminations (or disbonding), and impact damage. The first type is associated with the matrix and the second with both the matrix and the fibres.

Design and analysis

The ability to predict the propagation of damage in a structure is rather limited. It is however recognized that the criticality of damage depends on the level of strains in the structure.

Inspectability

Defects in composites range from small cracks and porosity to delaminations and impact damage. Specialized NDI methods are needed for proper assessment of composite components. At present there is no NDI method to evaluate bonding quality.

Experience and confidence level

Composite structures have not been in use for long enough to fully evaluate their long-term behaviour. It is therefore more likely that unexpected problems will arise.

Repair methods

The variety of defect and damage types in composite structures which may be coupled with several design concept calls for a relatively large number of repair methods. These methods require specialized equipment and training, and some of them are relatively complex. At present, the user may have to rely on manufacturers' assistance to deal with non-standard repairs.

IAI APPROACH

The design philosophy used for the composite components is aimed at having inspection-free structures. This means that the structure has to be able to function normally when containing manufacturing defects of allowable sizes, or after suffering impact which causes barely visible damage on the outer surface. The allowable manufacturing defect sizes are defined by analysis and the effect of defect coupon tests and are incorporated into engineering documentation.

Structures containing defects or damage, above the allowable levels, go to MRB (Material Review Board). The energy for barely visible damage is checked through impact tests. The damaged specimens are then tested to failure and their residual strength determined. This test data is used to define a strain allowable for the structure below which barely visible impact damage will not grow. Full scale tests are performed on major composite components. The test specimens contain implanted manufacturing defects and impact damage for final design substantiation. Some test specimens also undergo additional tests after being repaired in order to verify the validity of the repair procedures.

An effort is also being made to develop a means of enhanced visual detection of impact damage to the level at which each zone of the structure is designed to tolerate. Special repair methods are being developed for the different types of defects and damage; these repair methods have to supply the user with the ability to detect and repair the most common discrepancies in all the composite components of the aircraft. Repairs are divided into categories according to the ability of the agency which performs them.

IAI EXPERIENCE WITH THE LAVI

The Lavi is designed with all of its lift and control surfaces made of graphite/epoxy composites. Composites total about 25% of the structural weight. An important factor in supportability is the large area of composite surface.

A tight schedule during development often dictates the acceptance of components containing defects or damage of sizes in excess of the allowable limits without repair. This is possible for development aircraft, since the flight envelope is opened gradually. However, this calls for extensive use of Non-Destructive Inspection (NDI) at the initial stages, until enough confidence is obtained.

NDI is done as much as possible during grounding of the aircraft for scheduled maintenance. There is also some benefit in these inspections since the data gained can be used in future MRB decisions and in updating acceptance–rejection criteria.

As for in-service damage of composite structures, there were less than 10 cases to date on the Lavi prototype airplanes. Among them were:

—Damage to free edges of control surfaces.

—Heavy object falling on the canard surface which caused a slight dent. The damage occurred in a lightly loaded area and was repaired with a wet layup patch.
—Lower cover of a wing was damaged in front spar area when running hydraulic systems tests. The damage was repaired by a bolted metal doubler plate.
—A vertical fin was hit by a working platform and the skin near the edge was damaged. The fin was repaired by a wet lay-up patch.

These cases illustrate potential problems which may occur during routine service and they are covered in the repair procedures.

CONCLUSIONS AND RECOMMENDATIONS

Lack of toughness is the major problem in composite structures today. Tougher material systems should be utilized as well as methods for local toughening (e.g. soft interlayers). As far as maintenance is concerned, the user should be supplied with easy to apply repair methods (e.g. bolted titanium plate, wet lay-up patch). These methods should enable the user to be self-sufficient in solving standard repair problems. Appropriate tools should be available at all maintenance shops. Maintenance personnel should be specially trained with regular intensive retraining programs.

Summarizing our experience to date, we can state clearly, that the use of composites offers structural advantages. However, we do not have enough service experience to adequately support composite airframes. International cooperation in the exchange of experience and ideas is recommended for mutual benefit. As more experience is gained, new problems will arise, but finding solutions to these will be eased if such cooperation takes place. It is worthwhile considering ways in which to continue the initiative of this workshop and transform it into an on-going exchange of experience and ideas.

Composite Structures **10** (1988) 109–115

Supportability of Advanced Composite Structures

Thomas H. Bennett

Structures and Design Department Staff, General Dynamics/Fort Worth Division, Box 748 Fort Worth, Texas 76101, USA

ABSTRACT

Advanced composite airframe structures are entering service at an ever increasing rate. Early applications to secondary and stiffness-critical primary structures have been successful and have provided a wealth of operational support experience. This paper will offer an operator's definition of supportability as applied to advanced composite structures. Supportability issues and implications, identified from early service experience, will be discussed. Approaches will be offered for solution of advanced composite structures problems, before they occur.

INTRODUCTION

Advanced composite airframe structures have now been in service for more than a dozen years. The overall service experience has been good and the potential for weight reduction has been verified. Early applications were of unidirectional boron/epoxy or graphite/epoxy materials and most designs included aluminum honeycomb to form bonded assemblies. In most instances the composite parts have provided complete relief from the fatigue and corrosion problems associated with the aluminum parts they replaced. However, the operators have identified some unique advanced composite supportability issues, most of which could have been avoided or minimized, with little or no penalty, if they had been addressed early in the design process.

SUPPORTABILITY DEFINITION

Supportability of advanced composite structure from the operator's point of view is, 'The measure of all requirements necessary to retain unrestricted operation in any environment for the desired life of the structure.' This definition is broad and requires additional definition of several key words and phrases to help understand supportability from the operator's point of view.

The word 'measure' is applied to several terms normally associated with reliability, maintainability and logistics support. From the operator's point of view, the measures are relative values to be compared with current/recent experience. These comparisons include:

—number of people/skills;
—inspection/repair frequency;
—amount and complexity of support equipment;
—storage requirements for inspection/repair materials;
—time for inspection/repair;
—packaging/handling/transportation;
—down time;
—spares;
—facilities.

'Measures' can be affected by the operational scenario and will change with the usage. It is therefore important that the designer understands this and designs for the most severe conditions.

The definition of the phrase 'all requirements' is again broad from the operator's perspective. Essentially all the things measured in the above paragraph combine to make-up 'all requirements'. 'All requirements' include everything the operator's maintenance and support personnel must be, must have or must do. Again, these requirements change with the operational scenario.

The phrase 'unrestricted operation' includes not only flight but also other operational requirements. 'Unrestricted operation' means:

—no flight limitations;
—no mobility constraints;
—no survivability penalties;
—no vulnerability reduction;
—no degradation of sortie generation rates;
—no operations and support cost increases.

It is important to note that these terms apply to commercial and private aircraft operations as well as to military usage. In the past, operations and support costs have often been permitted to rise to 'make poor designs work'. Increasingly, operators are providing requirements which specifically prohibit this approach and in many cases require dramatic reductions in operations and support costs.

The term 'any environment' as used in this definition applies to both operational and climatic environment. The ability to support advanced composite airframe structure can vary dramatically depending on the climatic as well as the basing mode (environment from which the aircraft operates). The definition of 'any environment' can include:

—major overhaul facility/depot;
—rain forest/desert/arctic;
—major operating base/airport;
—dispersed operating location;
—private airfield;
—austere bases;
—any base under attack;
—air combat situations;
—long term storage;
—space.

All of the potential operational environments may not be known when the structures are being designed, which could result in limits to supportability. Fortunately the operators are becoming more proficient in identifying the operational conditions for the designer. Again the design must be supportable in the worst environment or the operator can be severely handicapped.

The phrase 'desired life of the structure' should be taken from the historical perspective. Operation to the originally specified usage and life is the exception and not the rule. 'Desired life of the structure' includes:

—originally specified life;
—originally specified usage;
—modified usage;
—extended life.

More of the aircraft are being adapted to unanticipated usage and the economic situation often dictates that the life be extended. The definition of 'desired life of the structure' again is the prerogative of the operator and historically changes over time. The designer cannot be expected to anticipate all eventual requirements, but must be aware of the potential restrictions which the material selection and design might impose. Aircraft which enjoy long production runs and many modifications are those which can economically accommodate changes to original requirements.

SUPPORTABILITY ISSUES

Supportability issues which have been identified from the service experience generally fall into three categories:

—damage susceptibility;
—inspection requirements;
—repair requirements.

Damage susceptibility

Damage susceptibility of the current generation of composite structures has proven to be a concern for the operator. The fragile nature of the early composite materials, combined with their application to thin skinned honeycomb structures resulted in more low velocity impact damage than was anticipated. Identified damage sources include:

—maintenance, drops/bumps;
—hail/lightning;
—foreign objects/birds.

Much of the damage to these early structures was caused by maintenance activities which caused no damage to aircraft with metallic structures. New toughened composite materials will improve resistance to maintenance activities and the maintenance procedures are being revised to reduce damage to composite structures. Even with these changes it is still important for the designer to understand the full range of damage sources, and to match the material and designs to all the damage threats.

Inspection requirements

Inspection requirements, as an issue for composite structures, is closely related to damage susceptibility. The random nature of the impact damage, and the potential for causing internal damage without leaving a surface indication, make inspection an issue for the operator. Inspection requirement issues include:

—inspection frequency;
—equipment/skills;
—assessment of inspection results;
—tracking flaw/damage growth;
—new materials/different failures.

As new composite materials are applied we must be sure that we understand, and can inspect for, the new failure modes at each level of support.

Repair requirements

Repair requirements such as the following have become issues to the operator:

—repair limits;
—frequency of repair;
—equipment/skills;
—repair materials;
—repair time;
—repair integrity.

Unfortunately, these requirements have often been thought of only after the design has been optimized for performance and manufacturing cost. Repair requirements and constraints should be determined up-front. Repair development should be based on the threats, and have repair limits/capabilities which match both the threats and operating conditions where the damage is most likely to occur. Validation of repair times, conditions and integrity should be part of the mainstream design development effort.

SUPPORTABILITY SOLUTIONS

Approaches to solving the potential composite supportability problems, before they become problems are as follows:

—know total operational scenario;
—design for supportability;
—understand composite support;
—develop support for composites.

The total operational scenario includes the following:

—flight loads/temperature;
—landing/taxi loads, FOD;
—hail/lightning/temperature;
—maintenance levels/threats;
—combat threats;
—support constraints.

We typically do a good job of anticipating flight and landing loads. Unfortunately, for advanced composite structures it is often the other, less understood, portions of the operational scenario which pose supportability problems. However, these portions can also be predicted, and designed for, once the need is understood.

The task of designing for supportability must include the following:

—consideration during research;
—materials characterization;
—conceptual design consideration;
—design allowables/consider repair;
—design to minimize damage;
—consider damage assessment;
—consider repair restraints;
—develop and test repairs.

The task of designing for supportability must start early. Both the research and conceptual design efforts should include supportability requirements/constraints. As new materials are introduced for consideration, the material allowables development should produce the data needed to assess supportability. Damage threats, inspection capabilities and repair constraints should be considerations when determining design allowables. Repairs should be designed and demonstrated as part of the design development effort.

An understanding of composite supportability includes:

—understanding the fragile nature of composites;
—know/avoid damage causes;
—special handling requirements;
—inspection/assessment capability;
—repair skill, equipment, material requirements;
—paint removal;
—corrosion potential.

It is equally important that the user and supporter understand that advanced composite structures require support different from that of metallic structures. Many of the damages to the early composite structures could have been prevented if the user and/or supporter had been aware of the damage susceptibility of these new materials.

Once the support requirements for advanced composite structures are understood, the operator can develop support for composites. Activities to develop support for composites should begin with research and include:

—training for operations and support personnel;

—development of assessment skills and equipment;
—development of repair skills, equipment and materials;
—damage from identified threats;
—repair limits at each level;
—repair times.

Many research activities are still needed to increase the supportability of advanced composite structures. This is especially true for new materials being introduced. It is also important that we develop the capability to match the damage potential to the threat, the damage assessment/repair capability to the operational support scenario and time requirements. We will never be able to avoid all problems through design and care, therefore new support capabilities must be developed specifically for composite structures.

CONCLUSIONS

Supportability of advanced composite structures is a complex subject involving many different perspectives. The most important perspective is that of the operator. The researcher and the designer can solve most of the composite supportability problems by knowing and addressing the potential problems early. This approach will be necessary for operator acceptance of the accelerated introduction of advanced composite airframe structures.

If we understand the total operational requirements early in the research and design efforts, conduct the necessary research and design for supportability and recognize that composites require changes in support and make those changes, supportability of advanced composite structures will not become a problem.

Index